ELECTROWEAK INTERACTIONS

ELECTROWEAK INTERACTIONS

Luciano Maiani

Università di Roma La Sapienza and INFN, Roma, Italy

Translated from the original Italian by Geoffrey Hall, Imperial College, London

CRC Press
Taylor & Francis Group
Boca Raton London New York

CRC Press is an imprint of the
Taylor & Francis Group, an **informa** business

This book was first published in Italian in 2013, by Editori Riuniti University Press, under the original title: Interazioni Elettrodeboli.

CRC Press
Taylor & Francis Group
6000 Broken Sound Parkway NW, Suite 300
Boca Raton, FL 33487-2742

First issued in paperback 2020

© 2016 by Taylor & Francis Group, LLC
CRC Press is an imprint of Taylor & Francis Group, an Informa business

No claim to original U.S. Government works

Version Date: 20151030

ISBN 13: 978-0-367-57522-9 (pbk)
ISBN 13: 978-1-4987-2225-4 (hbk)

Visit the Taylor & Francis Web site at
http://www.taylorandfrancis.com

and the CRC Press Web site at
http://www.crcpress.com

Contents

List of Figures

List of Tables

Preface

In the 2006/2007 academic year, the masters (*Laurea Magistrale*) courses were reorganised to match the lines of research in the physics department of the University of Rome "La Sapienza". The traditional course of theoretical physics, then focused on quantum electrodynamics with elements of renormalisation theory and some incursions into the standard electroweak theory, was divided into three single-term modules.

A first module (relativistic quantum mechanics) was to demonstrate the path from non-relativistic quantum mechanics to relativistic field theory, the fundamental paradigm of the modern theory of elementary particles and many other areas of physics, from condensed matter to astrophysics and cosmology. As such, the course was aimed at a general audience of those taking the masters degree course in physics.

The second module was expected to be dedicated to an introduction to the modern theory of elementary particles, in particular to the unified theory of the weak and electromagnetic interactions. The idea was to keep the course at an elementary level, so that it could be taken by theory students, as a phenomenological foundation for the more formal developments of the third module, and by experimental students, for completion of their theoretical training.

Finally, the third module was to provide an introduction to modern gauge theories, including renormalisation, with a decidedly theoretical style.

Further developments of a more formal nature were left to specialist courses, such as the course of field theory, traditionally taken in the fifth and final year of the masters course, or in doctoral courses.

The three modules proved very successful, in particular for training the new generation of researchers, both experimental and theoretical, who today are studying the physics of the Large Hadron Collider, LHC, at CERN.

The course lecture notes form the basis of a three–volume series, published originally by Editori Riuniti University Press. The plan took shape with the publication of the first volume, *Relativistic Quantum Mechanics: An Introduction to Relativistic Quantum Fields*, and now continues with the second volume, *Electroweak Interactions*. We hope soon to complete the third volume (*Introduction to Gauge Theories*), from the course taught in collaboration with Nicola Cabibbo, who also prepared a good fraction of the lecture notes.

Electroweak Interactions illustrates the conceptual route which led to unification of the weak and electromagnetic interactions starting from the identification of the weak hadronic current with the isotopic spin current, the CVC hypothesis of Feynman, Gell–Mann and others, and leading to Yang–Mills theory and to the first electroweak theory of Glashow. Particular attention is paid to the mechanism of spontaneous breaking of a global symmetry, with the Goldstone theorem, and of a local symmetry, with the Higgs–Brout–Englert mechanism, to arrive at the theory of Weinberg and Salam, based on the spontaneous breaking of the local $SU(2)_L \otimes U(1)_Y$ symmetry.

A brief introduction to the theory of quarks is followed by a description of quark mixing, the Cabibbo angle and the Glashow–Iliopoulos–Maiani mechanism, leading to the theory of Kobayashi and Maskawa with six quark flavours and CP violation.

The present volume, including appendices, provides the fundamental elements of the theory of compact Lie groups and their representations, necessary for a basic understanding of the role of flavour symmetry in particle physics.

The book contains a selection of phenomenological topics: neutral current interactions of neutrinos, determination of the physical parameters of the Z^0 in electron–positron reactions, decays of vector bosons, determination of the mass of the light quarks, phenomenology of CP violation in the neutral K-meson system. For these applications, the field theory ideas introduced in the first volume of the series [1] are sufficient and later a superficial acquaintance with Feynman diagrams which the reader can find, for example, in [2] or, in the future, in the third volume of this series [3].

A separate, more advanced, chapter is dedicated to strangeness violating, or more generally flavour changing, neutral current processes. Repeating what happened historically with the discovery of charm, the study of these processes has taken on a significant importance in recent years, as a 'probe' to reveal the presence of new phenomena at energies not yet accessible with particle accelerators.

In the final chapter, the expected properties of the Higgs–Brout–Englert boson and the methods adopted for its search are explained. The predictions are compared with relevant experimental results from the recent discovery, on the part of the ATLAS and CMS collaborations at CERN in July 2012, of a particle with a mass of around 125 GeV, which, at present, seems compatible with expectations of this key particle in the electroweak theory.

These lectures owe much to the collaboration over more than twenty years with Nicola Cabibbo and to feedback, over many years of teaching, from my students. Special thanks to Omar Benhar, Gino Isidori, Veronica Riquer and Massimo Testa, for comments and advice, and to Antonello Polosa for careful editing of the manuscript.

The Author

Luciano Maiani is emeritus professor of theoretical physics at the University of Rome, "La Sapienza", and author of more than two hundred scientific publications on the theoretical physics of elementary particles. He, together with Sheldon Glashow and John Iliopoulos, made the prediction of a new family of particles, those with 'charm', which form an essential part of the unified theory of the weak and electromagnetic forces. He has been president of the Italian Institute for Nuclear Physics (INFN), Director General of CERN in Geneva and president of the Italian National Council for Research (CNR). He promoted the development of the Virgo Observatory for gravitational wave detection, the neutrino beam from CERN to Gran Sasso and at CERN directed the crucial phases of the construction of the Large Hadron Collider. He has taught and worked in numerous foreign institutes. He was head of the theoretical physics department at the University of Rome, "La Sapienza", from 1976 to 1984 and held the chair of theoretical physics from 1984 to 2011. He is a member of the Italian Lincean Academy and a Fellow of the American Physical Society.

THE GENERAL PICTURE

CONTENTS

1.1 INTRODUCTION

The search for a theory which unifies the principles of quantum mechanics with the requirements of special relativity led [1, 2] to quantum field theory (QFT). The dynamic variables of the theory are *quantised fields*, linear operators on the Hilbert space, $\mathcal{H}$, of the states

$$\psi(x), A^\mu(x), \phi(x), \ldots$$

operators which depend on the space-time location and have definite transformation properties (spinors, 4-vectors, scalars, etc.) for changes from one reference system to another:

$$\psi'(x') = S(\Lambda)\psi(x), \ \ \text{etc.}$$

where Λ belongs to the proper Lorentz group.

Apart from pathological cases, the theory contains one state of minimum energy, invariant under translations, which we interpret as the *vacuum state*. This state is unique[1].

Microcausality. Starting from the fields, the local *observables*, $\mathcal{O}(x)$, are constructed, for example the energy density, charge density, etc. The condition of *microcausality* requires that different observables in different locations should commute with each other for space-like separations:

$$[\mathcal{O}(x), \mathcal{O}'(y)] = 0 \ \text{ if } \ (x-y)^2 < 0. \tag{1.1}$$

---

[1]Even in the case of an exact symmetry with spontaneous breaking, which leads to degenerate vacuum states, we can return to this situation by introducing a small external perturbation so as to prefer a single state, and let the perturbation tend to zero at the end of the calculation.

In its turn, equation (1.1) requires that the fields commute or anticommute among themselves for spatial separations:

$$[\psi(x), \psi(y)]_\pm = 0 \quad \text{if} \quad (x - y)^2 < 0 \tag{1.2}$$

where the $-$ (+) sign denotes the commutator (anticommutator) and applies to fields of integer (half-integer) spin.

The spin-statistics relation. In certain circumstances the fields can be expanded in quantised plane waves. The corresponding creation operators, applied to the vacuum state, give rise to states of a relativistic particle with a given mass, and with a spin defined by the the transformation properties of the field. The quanta of a given field are identical particles which conform to statistics characterised by the commutation or anticommutation rules respected by the corresponding field (Bose–Einstein or Fermi–Dirac, respectively). The spin-statistics theorem states that fields corresponding to quanta of integer (half-integer) spin have to be quantized with commutation (anticommutation) rules.

Antiparticles. The relativistic relation between the energy and momentum of a particle:

$$E^2 = \mathbf{p}^2 + m^2 \tag{1.3}$$

has solutions with both positive and negative energies. In QFT these solutions correspond to fields which evolve in time with positive or negative frequencies. For electrically charged fields, the corresponding creation operators produce, from the vacuum, particles with opposite electrical charges and equal mass and spin; the field quanta are the particles and corresponding *antiparticles*. Neutral particles which possess a conserved charge also have an antiparticle with an opposite value of the charge (an example is the antineutron, which has baryon number -1). Particles like the photon or π^0 meson which are their own antiparticles are known as intrinsically neutral.

Particles and antiparticles are connected by the CPT transformation, a combination of charge conjugation, parity inversion and time reversal. The CPT *theorem* guarantees that the mass, spin and, if applicable, the lifetime of a particle and its antiparticle are exactly equal.

The relation between spin and statistics, and the CPT theorem, are among the strongest arguments supporting QFT. It is a fact that no violations of these theorems have ever been observed, to a very stringent precision.

The masses of proton and antiproton coincide to within one part in 10 billion, while for the mass difference between the K^0 and its $\bar{K}^0$ antiparticle we find a limit of the order of $10^{-17} m_K$, as we will see in chapter 12; equations (12.31) and (12.32).

Fundamental degrees of freedom. The correspondence between a type of particle and a quantum field is a very delicate question, the first among the central problems of every QFT. A nucleus of the oxygen atom behaves like a pointlike particle in certain conditions; however we do not think of associating a quantum field to every atomic nucleus. How to decide, therefore, to which particles we should attribute a fundamental field? The answer is linked to the energy range with which we are able to observe the particle in question. The evidence we have today is that the leptons: electron, muon and τ and their corresponding neutrinos, are elementary, together with the photon and the mediators of the other fundamental forces. It is reasonable therefore to introduce a field for each one of these particles. On the contrary, the proton, neutron and other particles affected by nuclear interactions show themselves, already at energies available today, to be composed of more fundamental spin $\frac{1}{2}$ particles, the quarks, and spin 1 entities, the gluons.

The current QFT of the fundamental interactions, the so-called *Standard Theory*, is based on quarks, leptons and the mediators of the forces, as shown in Figure 1.1. A quantised field is associated with each of these particles.

Naturally, the situation could change if, at higher energy scales, one of these particles were to be shown in its turn to be composed of more fundamental constituents as hypothesised by several authors, such as *preons*, proposed by Pati and Salam, or *rishons*, introduced by Harari.

Dynamics. The field dynamics are generated from the Lagrangian density, $\mathcal{L}(x)$, a local function of the fields and their derivatives. In classical mechanics, the equation of motion of the fields is derived from $\mathcal{L}(x)$ by a *principle of least action*:

$$\delta \left(\int d^4x \, \mathcal{L}(x) \right) = 0, \quad \text{from which} \tag{1.4}$$

$$\partial^\mu \left(\frac{\partial \mathcal{L}}{\partial \partial_\mu \phi} \right) = \frac{\partial \mathcal{L}}{\partial \phi}. \tag{1.5}$$

From a quantum viewpoint, the fields are operators whose time evolution is given by the Heisenberg equations:

$$i\frac{\partial}{\partial t}\psi(x) = [\mathbf{H}, \psi(x)] \tag{1.6}$$

where $\mathbf{H}$ is the Hamiltonian derived from $\mathcal{L}$. Along with the canonical commutation (or anticommutation) relations between fields and conjugate momenta, equation (1.6) gives the Euler–Lagrange equations of motion (1.5).

In the presence of gauge fields, it is more advantageous to replace canonical quantisation with the path integral, which will be introduced in [3]. The starting point is the Green's functions of the fields:

$$G(x_1, x_2, \ldots, x_n) = <0|T\left(\phi(x_1)\phi(x_2)\ldots\phi(x_n)\right)|0> . \tag{1.7}$$

1. *Ordinary Matter (Stars, Earth, us,...):*

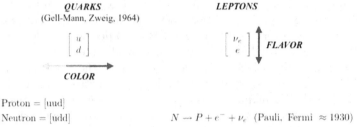

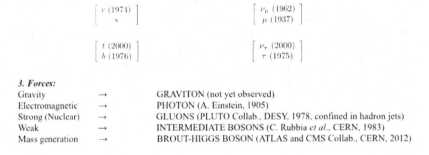

3. *Forces:*

Gravity	→	GRAVITON (not yet observed)
Electromagnetic	→	PHOTON (A. Einstein, 1905)
Strong (Nuclear)	→	GLUONS (PLUTO Collab., DESY, 1978, confined in hadron jets)
Weak	→	INTERMEDIATE BOSONS (C. Rubbia *et al.*, CERN, 1983)
Mass generation	→	BROUT-HIGGS BOSON (ATLAS and CMS Collab., CERN, 2012)

Figure 1.1 The fundamental constituents of matter and the forces. In the Standard Theory, a quantised field is associated with each particle. In brackets are the years in which the most recent constituents were first observed.

From the Green's functions, the S-matrix elements and particle lifetimes, etc. are obtained with the reduction formulae.

In general, the Lagrangian is written as the sum of a free part, defined by the fundamental properties of the particles, mass and spin, and an interaction part:

$$\mathcal{L} = \mathcal{L}_0 + \mathcal{L}_I. \tag{1.8}$$

Equation (1.8) presents us with two crucial problems of a QFT: how to write down the interaction Lagrangian? and how to calculate the physical amplitudes starting from $\mathcal{L}$?

In a certain sense the two problems are connected. The form of $\mathcal{L}_I$ must be modelled on experimental data, but this requires being able to switch from $\mathcal{L}_I$ to the data, or vice versa, to compare them with the predictions resulting from the working hypotheses we made about $\mathcal{L}_I$.

At the moment, trustworthy predictions can be obtained with *perturbation theory*, assuming that $\mathcal{L}_I$ is proportional to a small parameter, λ, and expanding the amplitudes in a power series in λ. The prototype of the method is quantum electrodynamics (QED), further evolved into the unified theory of the electroweak interactions which will be developed in this book. The predictive success of QED is still unprecedented.

The alternative, of treating the cases in which the interactions cannot be considered in any sense small, is numerical calculation of the quantum amplitudes in a discrete space-time, segmented into a finite mesh of points. The two methods are at present at different stages; simulation on a mesh has not yet reached the extraordinary precision of QED or even the more limited but still considerable level of the electroweak theory.

Fortunately, even for the strong, or nuclear, interactions, it is possible to identify an energy range in which perturbation theory can be applied, owing to so-called *asymptotic freedom*, and a careful comparison can be made of predictions derived from the Lagrangian with experimental data. This set of comparisons is the basis of our confidence in the Standard Theory, also including the strong interactions.

1.2 THE FORCES OF NATURE

We conclude this chapter with the characterisation of different components of $\mathcal{L}_I$. In nature four major classes of interaction well differentiated from each other have been observed, each with its own characteristics.

Gravitational interaction. The potential between two massive bodies is described by Newton's formula:

$$V(r) = G_N \frac{m_1 m_2}{r} \ . \tag{1.9}$$

As noticed long ago by Planck, in natural units Newton's constant, G_N, has the dimensions of the inverse of a mass squared. This mass is called the Planck mass and is extraordinarily large on the scale of atomic or nuclear masses:

$$M_P = G_N^{-1/2} = 1.22 \cdot 10^{+19} \text{ GeV} \ . \tag{1.10}$$

The effect is that, on the nuclear and atomic scale, gravitational interactions are extraordinarily weak and can be ignored altogether.

Electromagnetic interactions. Their strength is characterised by the *fine structure constant*:

$$\alpha = (\frac{e^2}{\hbar c}) \simeq \frac{1}{137} \tag{1.11}$$

where e is the absolute value of the electron charge. The small size of α allows significant use of perturbation theory. If we limit ourselves to charged leptons, the interaction Lagrangian is defined by the minimal substitution and we obtain QED [1]. The electromagnetic interactions of nuclear particles, collectively indicated with the term *hadrons*[2], are more complicated, given their non-elementary nature.

The electrostatic, Coulomb, force is *long range*:

$$V(r) = \frac{\alpha}{r} \; . \tag{1.12}$$

Given the vanishing mass of the photon, the Coulomb force dominates on distances of atomic scale. For larger distances, the strength of electrostatic forces is such that charges tend to neutralise one another and forces of the Van der Waals type, which result from molecular polarisation, dominate.

Strong or nuclear forces. This is the force which binds protons and neutrons together inside atomic nuclei, overcoming the electrostatic repulsion of the protons. The nuclear force has a short range, limited to the order of the nuclear radius:

$$V(r) \simeq \frac{e^{-r/R}}{r}$$
$$R \simeq 10^{-13} \text{ cm} = 1 \text{ fermi}. \tag{1.13}$$

A first theory of the nuclear interaction, proposed by Yukawa, described it as resulting from the exchange of particles similar to the photon, but having a mass: the π meson. In this case, the range of action is determined by the mass of the intermediate particle:

$$R \simeq \frac{\hbar}{m_\pi c} \; . \tag{1.14}$$

Comparing (1.14) with (1.13) we find[3] $m_\pi c^2 \simeq 100$–200 MeV, which is in agreement with the observed mass of the π meson discovered in cosmic rays shortly after the war, with $m_\pi c^2 \simeq 140$ MeV. The pion-nucleon coupling is characterised by a dimensionless constant analogous to the fine structure constant, but about 1000 times larger; the nuclear forces dominate over the electrostatic repulsion of protons in the nucleus. On atomic distances, however, the nuclear forces are completely irrelevant, owing to their exponential fall off, as in (1.13).

Weak interactions. These are the interactions responsible for neutron decay, discussed in [1]. The force is characterised by a dimensionless constant:

$$G_F m_p^2 \simeq 10^{-5} \tag{1.15}$$

where G_F is the Fermi constant, obtained from the average lifetime of the neutron [1]. The strength of the weak force is typically $(G_F m_p^2)^2/\alpha \simeq 10^{-7}$ that

---

[2]from the Greek word *hadros* = strong.
[3]The numerical result is obtained recalling that $\hbar \cdot c \simeq 197$ MeV $\cdot$ fermi.

of the electromagnetic force. The effects of the weak interaction on charged particles are observable only in the case of decays of particles which would otherwise be stable (for example the neutron) or in interactions of neutrinos (which are not sensitive to electromagnetic and nuclear forces).

Lifetimes of unstable particles The strong interactions allow the formation of *resonances*: unstable states which disintegrate into final particles as a result of the interaction. Typical decay times are of the order of the time needed for light to cross the resonance, which has a linear dimension of order R, therefore:

$$\tau \simeq \frac{R}{c} \simeq 10^{-23} \text{ s} \simeq \frac{\hbar}{m_\pi c^2} \, . \tag{1.16}$$

Times of this order of magnitude are measured by the *width*, the uncertainty in energy, of the resonance which is connected to the mean lifetime (1.16) by the uncertainty relation:

$$\Gamma \cdot \tau = \hbar \tag{1.17}$$

from which a typical width is obtained:

$$\Gamma \simeq m_\pi c^2 \simeq 100\text{--}200 \text{ MeV}. \tag{1.18}$$

For comparison, the typical average lifetimes for decays due to electromagnetic interactions (for example $\pi^0 \to \gamma\gamma$) are of order:

$$\tau_{e.m.} \simeq 10^{-16}\text{--}10^{-18} \text{ s}. \tag{1.19}$$

The rate of decay of the neutron is a very sensitive function of the n–p mass difference:

$$\Gamma_n = \frac{\hbar}{\tau_n} = \text{constant} \cdot (G_F)^2 (\Delta m)^5 \, . \tag{1.20}$$

The power with which Δm appears is determined by dimensional considerations, given that G_F in natural units has dimensions of $[mass]^{-2}$. Therefore for all weak decays we can set, at least as an order of magnitude:

$$\tau_X = \tau_n \left(\frac{\Delta m_{n-p}}{\Delta m_X}\right)^5 \tag{1.21}$$

where Δm_X is the mass difference between the X particle and those in the final state. In this way we find average lifetimes in the range from μs (for the muon, $\Delta m_\mu \simeq m_\mu = 110$ MeV) to 10^{-12} s (for particles containing a charmed quark with $\Delta m_c \simeq m_c \simeq 1.5$ GeV).

In any case, the separation between the orders of magnitude of the average lifetimes is notable; the lifetime of an unstable particle is a good indicator of the type of interaction which caused the decay.

In conclusion, the preceding considerations indicate an interaction Lagrangian of the form:

$$\mathcal{L}_I = \mathcal{L}_{strong} + \mathcal{L}_{e.m.} + \mathcal{L}_{weak}. \tag{1.22}$$

The form of the different terms is determined by the symmetries of the different interactions. In the Standard Theory which we will consider in what follows, these are all symmetries under *non-abelian gauge transformations*, generalisations of the familiar gauge transformation symmetry of the second type from classical and quantum electrodynamics.

This common denominator of interactions which are so different from the phenomenological point of view encourages us to think that the Standard Theory could be the manifestation of a structure still more symmetric and unified.

ISOTOPIC SPIN AND STRANGENESS

CONTENTS

The starting point is the observation that the proton and neutron have masses very close to each other:

$$\frac{m_n - m_p}{m_p} \simeq 0.15 \cdot 10^{-2} \ . \tag{2.1}$$

Heisenberg interpreted this fact as a consequence of the invariance of the strong interaction Hamiltonian under transformations which change the states of the proton and neutron into arbitrary complex superpositions:

$$\begin{pmatrix} p \\ n \end{pmatrix} \to U \begin{pmatrix} p \\ n \end{pmatrix} \tag{2.2}$$

where U is a unitary matrix, to preserve the normalisation $|p|^2 + |n|^2$. Every matrix of this type can be written as the product of a phase factor and a *special* matrix, i.e. a matrix with determinant equal to unity:

$$U = e^{i\phi} U_{spec} \ .$$

The invariance of the Hamiltonian under transformations of the fields of the proton and neutron by a common phase factor leads (Noether's theorem) to the conservation of *baryon current* and *baryon number*[1]:

$$J_B^\mu = \bar{p}\gamma^\mu p + \bar{n}\gamma^\mu n + \ldots; \qquad \partial_\mu J_B^\mu = 0;$$

$$N_B = \int d^3x\, J_B^0 = N_p - N_{\bar{p}} + N_n - N_{\bar{n}} + \ldots; \qquad \frac{dN_B}{dt} = 0 \ . \qquad (2.3)$$

Conservation of baryon number can be considered separately, therefore from now onwards we will restrict ourselves to transformations of the type (2.2) with:

$$det(U) = 1 \ . \qquad (2.4)$$

Mathematically, the transformations (2.2) with the condition (2.4) define the $SU(2)$ group of special unitary matrices in two dimensions. This group is locally isomorphous, in an infinitesimal region around unity, to the group of rotations in a three-dimensional Euclidean space, $O(3)$. $SU(2)$ is the group which represents spatial rotations on quantum states, allowing the inclusion of states of half-integer angular momentum (cf. [1]).

2.1 ISOTOPIC SPIN

From a physical point of view, given the symmetry under (2.2) and (2.4), we can consider the proton and neutron as the two charge states of a single particle, the *nucleon*. The analogy with the two spin states of a spin $\frac{1}{2}$ particle is clear, hence the name *isotopic spin*, or just isospin, given the symmetry under the transformations (2.2).

As we underlined in the previous chapter, the proton and neutron do not have a special role in the panorama of hadrons. It is convenient to interpret the symmetry (2.2) in a different way. Let us imagine that there exists a group G of transformations of the fundamental degrees of freedom of the strong interactions (quarks) which leave the Hamiltonian invariant. Given a transformation $g \in G$, the effect of g on the p and n states is actually of the form (2.2), with U obviously a function of g:

$$\begin{pmatrix} p \\ n \end{pmatrix} \rightarrow U(g) \begin{pmatrix} p \\ n \end{pmatrix} \ . \qquad (2.5)$$

If we carry out two successive transformations, first g_1 and then g_2, the effect will be to obtain the product transformation of the two: $g = g_2 \cdot g_1$. Correspondingly, we expect:

$$U(g_2) \cdot U(g_1) = U(g_2 \cdot g_1) \ . \qquad (2.6)$$

The matrices $U(g)$ provide a *representation* of the transformations of the

---

[1]The ellipsis dots denote that, in general, it is necessary to add to (2.3) the contributions of other particles which decay into mesons plus p or n and therefore they also must be transformed to respect the current conservation.

group G, in the mathematical sense of the term; the law of multiplication of the group is represented by the product of the matrices U (we have already encountered this concept in the context of the Lorentz transformations [1]; cf. Appendix C for a brief summary of group theory and their representations).

The matrices $U(g)$ themselves form a non-commutative group. Because the simplest non-commutative group is actually $SU(2)$, we conclude that the fundamental invariance group G must contain at least one $SU(2)$ subgroup which is represented in (2.5).

From the theory of angular momentum, we know that the vectors of an irreducible representation of $SU(2)$ are characterised by the total angular momentum and the eigenvector of its third component. In the case of isotopic spin, we denote the corresponding quantities by I and I_3. The dimension of the representation with isotopic spin I is equal to $2I + 1$. Therefore the nucleons have isotopic spin $\frac{1}{2}$.

The transformation (2.5) can now be extended to all hadronic states, which will generally transform according to representations different from that of the nucleons. The first example is provided by the Yukawa mesons. The carriers of the nuclear forces appear in a *triplet* of almost degenerate states, π^+, π^0, π^-, which we interpret as three charged states of a single particle, the π meson, or pion, which therefore has $I = 1$.

That hadrons appear in charged multiplets approximately degenerate in mass is one of the best respected predictions in particle physics (cf. Figures 2.1 to 2.4). In Figures 2.1 and 2.2 antiparticle masses are not indicated as they are equal to the corresponding particle mass within errors, as required by CPT symmetry, e.g. $M(\pi^-) = M(\pi^+)$.

The isotopic spin generators. In the Hilbert space of quantum states, the isospin transformations are represented by infinite-dimensional operators. For each $g \, \epsilon \, SU(2)$:

$$g \to \mathcal{U}(g); \qquad \mathcal{U}(g)^\dagger \mathcal{U}(g) = 1 \qquad (2.7)$$

and for each state $|A> \, \epsilon \, \mathcal{H}$:

$$|A> \to \mathcal{U}(g)|A>$$
$$\mathcal{U}(g) \, \mathbf{H}_{strong} \, \mathcal{U}(g)^\dagger = \mathbf{H}_{strong} \, . \qquad (2.8)$$

In an infinitesimal region around the identity:

$$\mathcal{U}(g) = 1 + \sum_{i=1,3} \alpha_i \mathcal{I}^i \qquad (2.9)$$

with $\mathcal{I}^i$ Hermitian operators which commute with the strong Hamiltonian, if we neglect mass differences like Δm_{p-n}, and that satisfy the commutation relations of the $SU(2)$ algebra :

$$[\mathcal{I}^i, \mathcal{I}^j] = i \, \epsilon^{ijk} \, \mathcal{I}^k \, . \qquad (2.10)$$

The $SU(2)$ representation provided by the $\mathcal{U}$ matrices, and also by the $\mathcal{I}^i$, is *completely reducible* into finite-dimensional representations (cf. Appendix C), and is constituted by matrices divided into finite-dimensional diagonal blocks, which correspond to isospin multiplets. We limit ourselves to single particle states, which we write as:

$$|\mathbf{p}, s, h, i, i_3, \alpha > \tag{2.11}$$

with $\mathbf{p}$ the momentum, s and h the spin and helicity[2], and α the other quantum numbers (B and S). Because $SU(2)$ commutes with the Hamiltonian, the transformations g change the state (2.11) into states of the same multiplet and we therefore have:

$$\mathcal{U}(g)|\mathbf{p}, s, h, i, i_3, \alpha > = \sum_{i_3'} U^{(i)}(g)_{i_3, i_3'}|\mathbf{p}, s, h, i, i_3', \alpha > \tag{2.12}$$

where $U^{(i)}(g)$ are the matrices of the representation with isospin i.

Correspondingly:

$$\mathcal{I}^k|\mathbf{p}, s, h, i, i_3, \alpha > = \sum_{i_3'} (I^{(i)})^k_{i_3, i_3'}|\mathbf{p}, s, h, i, i_3', \alpha > \tag{2.13}$$

where the $I^{(i)}_{i_3, i_3'}$ matrices have the same form as the matrices corresponding to angular momentum i, given, e.g., in [4].

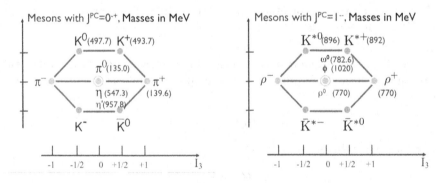

Figure 2.1   The light pseudoscalar mesons.   Figure 2.2   The light vector mesons.

2.2   STRANGENESS

The hadronic particles possess a further conserved quantum number, *strangeness*[3], S. Conservation of S in the strong interaction explains the long lifetime of several hadrons (for example K mesons or the Λ and Σ baryons)

---

[2]The spin component in the direction of $\mathbf{p}$.
[3]Introduced by Gell–Mann in the 1950s.

Table 2.1 Decays of light pseudoscalar mesons. Note the difference in lifetimes between the weak and electromagnetic decays.

	S	Dominant decay	ΔI	$|\Delta S|$	$\tau(s)$	Interaction
$\pi^{\pm}$	0	$\mu \, \nu_{\mu}$	1	0	$2.6 \cdot 10^{-8}$	weak
π^0	0	$\gamma \, \gamma$	1	0	$8.4 \cdot 10^{-17}$	electromagnetic
$K^{\pm}$	± 1	$\pi^{\pm} \, \pi^0$ $\mu \, \nu_{\mu}$ $\pi \, l \, \nu_l$	$\frac{3}{2}$ $\frac{1}{2}$ $\frac{1}{2}$	1	$1.2 \cdot 10^{-8}$	weak
K_L	± 1	3π $\pi \, l \, \nu_l$	$\frac{1}{2}, \frac{3}{2}$ $\frac{1}{2}$	1	$5.2 \cdot 10^{-8}$	weak
K_S	± 1	2π	$\frac{1}{2}$	1	$0.89 \cdot 10^{-10}$	weak
η	0	3π $\gamma \, \gamma$	≥ 1	0	$0.55 \cdot 10^{-18}$	electromagnetic

that, being the lightest particles with $S \neq 0$, can decay only through the weak interaction, with lifetimes of order 10^{-8}–10^{-12} sec.

We can assign zero strangeness to the nucleon and pion, and an appropriate strangeness value to every other hadron so that S is conserved in reaction and decay products of the strong interaction, and is violated, where applicable, in weak decays. Together with quantum numbers linked to Lorentz transformations, mass and spin, *isotopic spin, strangeness and baryon number* make up a complete system of quantum numbers, capable of characterising all light hadrons (i.e. with mass less than about 2 GeV). The electric charge can be obtained from the relation discovered empirically by Gell–Mann and Nishijima:

$$Q = I_3 + \frac{B + S}{2} \; . \tag{2.14}$$

We see from (2.14) that the electric charge inside an isospin multiplet changes in steps of 1 with I_3, and furthermore that strangeness is also conserved by the electromagnetic interaction, which conserves I_3 and B.

It is convenient instead of S to introduce the so-called *hypercharge*, Y, equal to twice the average electric charge of the multiplet, $< Q >$. From (2.14) we find:

$$< Q >= \frac{S + B}{2} \tag{2.15}$$

because $< I_3 >= 0$, therefore we can also write:

$$Q = I_3 + \frac{1}{2}Y; \qquad Y = S + B . \qquad (2.16)$$

Figures 2.1–2.4 show the lightest hadrons for given values of spin, baryon number and intrinsic parity. Each hadron is represented by a point in the plane of the two characteristic quantum numbers, I_3 and S. The masses of the different particles are also indicated; the quasi-degeneracy of the isospin multiplets is clear.

In Tables 2.1–2.3 the lifetimes and principal decay modes of the same particles are listed. The selection rules respected by the different interactions are clearly reflected in the lifetime values.

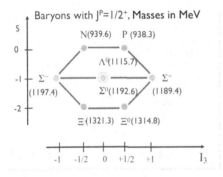

Figure 2.3   Spin $\frac{1}{2}$ baryons.

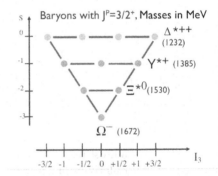

Figure 2.4   Spin $\frac{3}{2}$ baryons.

2.3   PION-NUCLEON AMPLITUDES

As well as the masses, isospin symmetry has as a consequence the existence of measurable relations between scattering amplitudes. We consider the simplest case of pion-nucleon scattering.

We describe the pion-nucleon states with the tensor product:

$$|\pi, m > |N, l >= |\pi, m; N, l > \qquad (2.17)$$

where we have indicated the I_3 quantum numbers of pions and nucleons, $m = \pm 1, 0$ and $l = \pm \frac{1}{2}$, and we have omitted the dependence on the particle momenta and spin, which stay fixed throughout the calculation. The scattering amplitudes are the S-matrix elements:

$$< \pi, m'; N, l'|S|\pi, m; N, l > . \qquad (2.18)$$

We can express the states (2.17) in terms of the eigenstates of total isospin and its third component, using the familiar Clebsch–Gordan coefficients relevant to the angular momentum composition, in this case $1 \otimes 1/2$:

$$|\pi, m; N, l> = \sum_{I=1/2,3/2} C(1, m; 1/2, l \mid I, m+l)|\pi, N; I, m+l>. \qquad (2.19)$$

Naturally, in place of the matrix elements (2.18) we can express the observable quantities in terms of:

$$< \pi, N; I', I_3'|S|\pi, N; I, I_3 > = \delta_{I,I'}\delta_{I_3,I_3'}S(I, I_3). \qquad (2.20)$$

In this relation, we have already partially taken account of isospin symmetry; because S commutes with I and I_3 it must be diagonal in the basis in which these quantum numbers are diagonal. However, the symmetry implies that S commutes with *all the components* of isotopic spin, i.e. also with I_1 and I_2. An operator which satisfies these conditions in an irreducible representation must be a multiple of the identity matrix (this is the conclusion of *Schur's lemma*, cf. Appendix C). Therefore:

$$S(I, I_3) = \text{independent of } I_3 = S(I) . \qquad (2.21)$$

The pion-nucleon scattering amplitudes are determined by only two invariant amplitudes: $S(1/2)$ and $S(3/2)$.

The accessible experimental reactions involve the scattering of $\pi^{\pm}$ by a proton:

$$\pi^+ + p \to \pi^+ + p; \qquad (2.22)$$
$$\pi^- + p \to \pi^- + p; \qquad (2.23)$$
$$\pi^- + p \to \pi^0 + n \ \text{(charge exchange)}. \qquad (2.24)$$

Using numerical values of the Clebsch–Gordan coefficients [5], we find:

$$|\pi^+ p> = |3/2, +3/2>; \qquad (2.25)$$

$$|\pi^0 p> = \sqrt{\frac{2}{3}}|3/2, +1/2> - \sqrt{\frac{1}{3}}|1/2, +1/2>;$$
$$|\pi^+ n> = \sqrt{\frac{1}{3}}|3/2, +1/2> + \sqrt{\frac{2}{3}}|1/2, +1/2>; \qquad (2.26)$$

$$|\pi^- p> = \sqrt{\frac{1}{3}}|3/2, -1/2> - \sqrt{\frac{2}{3}}|1/2, -1/2>;$$
$$|\pi^0 n> = \sqrt{\frac{2}{3}}|3/2, -1/2> + \sqrt{\frac{1}{3}}|1/2, -1/2>; \qquad (2.27)$$

Table 2.2   Decays of baryons stable for strong interactions. For each decay mode, the branching fraction is indicated in parentheses. In the last column, the interaction responsible for the decay: w.=weak, e.m.=electromagnetic.

	S	Dominant decay	ΔI	$|\Delta S|$	$\tau(s)$	Int.
p	0	not observed			$\geq 10^{31-33}$ years	??
n	0	$p\,e^-\,\bar{\nu}_e$ (1)	1	0	885.7 ± 0.8	w.
Σ^+	-1	$p\,\pi^0$ (0.52) $n\,\pi^+$ (0.48) $\Lambda\,e^+\nu_e$ $(2.0 \pm 0.5\ \cdot 10^{-5})$	$\frac{1}{2}$ $\frac{1}{2}$ 1	1 1 0	$0.802 \pm 0.003 \cdot 10^{-10}$	w.
Σ^-	-1	$n\,\pi^-$ (0.998) $n\,e^+\bar{\nu}_e$ $(1.017 \pm 0.034\ \cdot 10^{-3})$ $\Lambda\,e^-\bar{\nu}_e$ $(5.73 \pm 0.27\ \cdot 10^{-3})$	$\frac{1}{2}$ $\frac{1}{2}$ 1	1 1 0	$1.479 \pm 0.011 \cdot 10^{-10}$	w.
Σ^0	-1	$\Lambda\,\gamma$ (1.00)	1	0	$7.4 \pm 0.7 \cdot 10^{-20}$	e.m.
Λ	-1	$p\,\pi^-$ (0.639) $n\,\pi^0$ (0.358) $p\,e^-\bar{\nu}_e$ $(8.32 \pm 0.14\ \cdot 10^{-4})$	$\frac{1}{2}$	1	$2.632 \pm 0.020 \cdot 10^{-10}$	w.
Ξ^0	-1	$\Lambda\,\pi^0$ (0.99522 ± 0.00032) $\Sigma^+ e^-\bar{\nu}_e$ $(2.7 \pm 0.4\ \cdot 10^{-4})$	$\frac{1}{2}$	1	$2.90 \pm 0.009 \cdot 10^{-10}$	w.
Ξ^-	-1	$\Lambda\,\pi^-$ (0.99887 ± 0.00035) $\Sigma^0 e^-\bar{\nu}_e$ $(8.7 \pm 1.7\ \cdot 10^{-5})$ $\Lambda\,e^-\bar{n}u_e$ $(5.63 \pm 0.3\ \cdot 10^{-4})$	$\frac{1}{2}$	1	$1.639 \pm 0.015 \cdot 10^{-10}$	w.

Table 2.3 Decays of the first hadronic resonances. Note the conservation of strangeness in all the decays except the last, for which the lifetime indicates a weak decay.

	Dominant decay	Γ (MeV)	Interaction
ρ	2π	149	strong
K^*	$K\,\pi$	51	strong
ω	3π	8.4	strong
ϕ	$K\,\bar{K}$	4.3	strong
Δ	$N\,\pi$	120	strong
Y^*	$\Lambda\,\pi$ $\Sigma\,\pi$	3.6	strong
Ξ^*	$\Xi\,\pi$	9	strong
Ω	$\Lambda\,K$ $\Xi\,\pi$	$0.8 \cdot 10^{-10}$ sec	weak

from which we easily find:

$$A(\pi^+ p \to \pi^+ p) = S(3/2)$$

$$A(\pi^- p \to \pi^- p) = \frac{1}{3}S(3/2) + \frac{2}{3}S(1/2)$$

$$A(\pi^- p \to \pi^0 n) = \frac{\sqrt{2}}{3}[S(3/2) - S(1/2)] . \tag{2.28}$$

For pion momenta greater than 200 MeV, the cross section increases rapidly, Figure 2.5. The effect is caused by a very prominent resonance which is present in the scattering of positive pions on the proton, and therefore must have isotopic spin $\frac{3}{2}$[4]. At the peak of the resonance we can neglect $S(1/2)$ in equation (2.28). Taking the square of the amplitudes, we find that the cross sections are in the ratios:

$$\sigma(\pi^+ p) : \sigma(\pi^- p)_{el} : \sigma(\pi^- p)_{ch.ex} = 9 : 1 : 2 . \tag{2.29}$$

The ratios (2.29) are very well satisfied by experimental values[5] at the

---

[4] Analysis of the angular distribution shows that the resonance has angular momentum $J = \frac{3}{2}$ and is therefore denoted with the notation $\Delta(\frac{3}{2}, \frac{3}{2})$. The first observation of the $\Delta(\frac{3}{2}, \frac{3}{2})$ was by Fermi at the Chicago cyclotron, one of the first accelerators dedicated to the study of elementary particles.

[5] The cross sections are given in *millibarns*, where 1 mb = 10^{-27} cm^2.

peak, (cf. Figure 2.5):

$$\sigma(\pi^+ p) \simeq 200 \text{ mb};$$
$$\sigma(\pi^- p)_{tot} = \sigma(\pi^- p)_{el} + \sigma(\pi^- p)_{ch.ex} \simeq 70 \text{ mb}. \qquad (2.30)$$

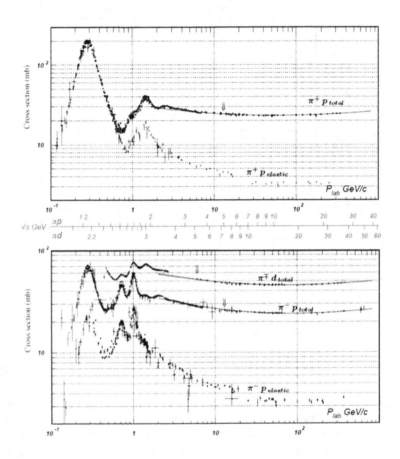

Figure 2.5 Cross sections for the scattering of pions on nuclei [5].

2.4 ISOSPIN AND WEAK CURRENTS

At not too high energies (typically, $E \leq 1$ GeV) a good approximation of hadronic physics can be obtained from a Lagrangian based on the fields of the nucleon, pion and other particles discussed in the previous section[6].

***

[6]At these energies the fundamental degrees of freedom are *frozen* inside the hadrons, which behave as if they were elementary particles.

Naturally, the strong interaction Lagrangian must be invariant under isotopic spin transformations and we can write it as:

$$\mathcal{L} = \bar{N}(i\partial\!\!\!/ - M)N + \frac{1}{2}(\partial_\mu \pi^i \partial^\mu \pi^i - m^2 \, \pi^i \pi^i) + g_\pi \bar{N} i\gamma_5 \tau^i N \cdot \pi^i + \ldots . \quad (2.31)$$

τ^i are the Pauli matrices, N and π^i denote the fields of the nucleon and the pion, respectively, M and m are their masses, the ellipsis dots denote the contributions from other fields and summation over repeated indices is understood. We have introduced an interaction term in the simplest form (Yukawa interaction) taking account of the negative parity of the pion (for which reason the γ_5 matrix is included).

The infinitesimal $SU(2)$ transformations for nucleons and pions are written explicitly as (cf. Appendix C):

$$\delta N = -i\alpha_i \cdot (I^{(1/2)})^i \, N = -i\alpha_i \cdot \frac{\tau^i}{2} N;$$

$$\delta \pi^k = -i \, \alpha_i [(I^{(1)})^i \, \pi]^k = \alpha_i \epsilon^{kij} \pi^j; \qquad (i, j, k = 1, 2, 3). \quad (2.32)$$

The pion transforms like the regular representation of $SU(2)$, whose generators are fixed by the $SU(2)$ structure constants.

For the pion, we have used real fields, $\pi^{1,2,3}$ (Cartesian representation), connected to the charged fields according to:

$$\pi^0 = \pi^3; \qquad \pi = \frac{\pi^1 + i\pi^2}{\sqrt{2}} . \quad (2.33)$$

The field π *creates* a π^+ and *destroys* its antiparticle, π^-.

Isotopic spin invariance gives rise to three conserved currents which, according to Noether's theorem, are written in the following way:

$$J_\mu^k = -i \sum_{campi} \frac{\partial \mathcal{L}}{\partial \, \partial^\mu \psi_\alpha} (I^k)_{\alpha,\beta} \psi_\beta \quad (2.34)$$

where the sum extends to all the fields in the Lagrangian, α, β are indices which identify the isospin components of the generic field ψ, and $(I^k)_{\alpha,\beta}$ are the matrices of the isospin generators in the multiplet of ψ. Calculating explicitly with (2.32), we find:

$$J_\mu^k = \bar{N}\gamma_\mu \frac{\tau^k}{2} N + (\partial_\mu \pi)^l \, \epsilon^{lks} \, \pi^s + \ldots \quad (2.35)$$

The $SU(2)$ generators defined in (2.9) and (2.10) are obtained from the currents in the usual way:

$$\mathcal{I}^k = \int d^3x \, J_0^k = \int d^3x \, [N^\dagger \frac{\tau^k}{2} N + (\partial_0 \pi)^l \, \epsilon^{lks} \, \pi^s] + \ldots \quad (2.36)$$

We leave it to the reader to demonstrate that the operators (2.36) satisfy the commutation rules of the algebra, equation (2.10), as a consequence of the canonical commutation and anticommutation rules of the π and N fields.

Now we calculate the current which corresponds to the *raising* operator of isotopic spin:

$$
\begin{aligned}
J_\mu^{(+)} = J_\mu^1 + iJ_\mu^2 = & \\
= & \, \bar{p}\gamma_\mu n + (\partial_\mu \pi^3)\pi^2 - (\partial_\mu \pi^2)\pi^3 + i[(\partial_\mu \pi^1)\pi^3 - (\partial_\mu \pi^3)\pi^1] = \\
= & \, \bar{p}\gamma_\mu n + [(\partial_\mu \pi^3)(\pi^2 - i\pi^1) - \pi^3 \partial_\mu(\pi^2 - \pi^1)] = \\
= & \, \bar{p}\gamma_\mu n - i\sqrt{2}[(\partial_\mu \pi^3)\frac{(\pi^1 + i\pi^2)}{\sqrt{2}} - \pi^3 \partial_\mu \frac{(\pi^1 + i\pi^2)}{\sqrt{2}}] = \\
= & \, \bar{p}\gamma_\mu n - i\sqrt{2}(\partial_\mu \pi^3 \pi - \pi^3 \partial_\mu \pi) \ .
\end{aligned}
\tag{2.37}
$$

The two terms of the current correspond to the transitions $n \to p$ and $\pi^- \to \pi^0$, respectively. The first term is precisely the vector current which enters into β decay of the neutron, which therefore coincides with the isotopic spin current of the nucleon.

Following Feynman and Gell–Mann, we make the hypothesis that the hadronic vector current which enters into the weak Lagrangian *is the same as the isotopic spin current* required by Noether's theorem and conserved as a result of that symmetry. This conjecture, known as the *conserved vector current* (CVC) hypothesis, reproduces the Fermi theory of the neutron, but gives us also the exact relative normalisation between the Fermi theory and the weak current which describes the β decay of the pion:

$$
\pi^- \to \pi^0 \ e^- \ \bar{\nu} \ .
\tag{2.38}
$$

We write the weak Lagrangian by comparison with neutron decay as (cf. [1]):

$$
\mathcal{L}_{weak}^{(n)} = \frac{G_F}{\sqrt{2}} \left[\bar{p}\gamma_m u(1 + \frac{g_A}{g_V}\gamma_5)n \right] [\bar{e}\gamma_m u(1 - \gamma_5)\nu_e] \ .
\tag{2.39}
$$

Then, according to the CVC hypothesis, the weak Lagrangian that describes the decay (2.38) must be:

$$
\mathcal{L}_{weak}^{(\pi)} = \frac{G_F}{\sqrt{2}} \left\{ -i\sqrt{2}[(\partial_\mu \pi^0)\pi^\dagger - \pi^0(\partial_\mu \pi^\dagger)][\bar{e}\gamma_m u(1 - \gamma_5)\nu_e] \right\}
\tag{2.40}
$$

where G_F is the same Fermi constant which determines neutron β decay.

Notice that there are no arbitrary constants in (2.40); the relative normalisation between the vector current of the neutron and that of the π^- is fixed by the normalisation of the respective generators; see (2.37). We can use (2.40) to *calculate* the β decay rate of the π^-. The result is in excellent agreement with the observed value.

Problem 1.   Calculate the decay rate of the pion in (2.38) starting from the Lagrangian (2.40).

Solution.   We denote with $p^\mu_{i,f} = (E_{i,f}, \mathbf{p}_{i,f})$ the initial and final pion momenta.

$$S_{fi} = (2\pi)^4 \delta^{(4)}(P_{out} - P_{in})(\sqrt{\frac{1}{V^2 4 E_i E_f}} \sqrt{\frac{m_e m_\nu}{V^2 E_e E_\nu}}) M_{fi};$$

$$M_{fi} = \frac{G_F}{\sqrt{2}} < f|J^\pi_\mu|i > l_\mu;$$

$$< f|J^{(\pi)}_\mu|i > = \frac{(p_i + p_f)_\mu}{2\sqrt{E_i E_f}} < I^{(+)} >; \qquad < I^{(+)} > = \sqrt{2}$$

$$l_\mu = \bar{u}_e \gamma_\mu (1 - \gamma_5) v_\nu. \tag{2.41}$$

$$d\Gamma = \frac{1}{2}\Sigma_{spin} \frac{|S_{fi}|^2}{T} \frac{V d^3 p_f}{(2\pi)^3} \frac{V d^3 p_e}{(2\pi)^3} \frac{V d^3 p_\nu}{(2\pi)^3} =$$

$$= (2\pi)^4 \delta^{(4)}(P_{out} - P_{in}) \left(\frac{G_F}{\sqrt{2}}\right)^2| < I^{(+)} > |^2 \frac{(p_i + p_f)_\mu}{2\sqrt{E_i E_f}} \frac{(p_i + p_f)_\nu}{2\sqrt{E_i E_f}} \times$$

$$\left(\frac{m_e m_\nu}{E_e E_\nu}\right) Tr[\frac{\not{e}\gamma^\mu(1 - \gamma_5)\not{\psi}\gamma^\nu(1 - \gamma_5)}{4 m_e m_\nu}] \frac{d^3 p_f}{(2\pi)^3} \frac{d^3 p_e}{(2\pi)^3} \frac{d^3 p_\nu}{(2\pi)^3}. \tag{2.42}$$

A few observations:

- There is a factor 2 which comes from the relation:

$$(1 - \gamma_5)^2 = 2(1 - \gamma_5). \tag{2.43}$$

This factor exactly cancels the factor $(1/\sqrt{2})^2$ which comes from the Lagrangian; the Fermi constant in the V–A theory is defined so as to obtain the same numeric value as in the simple Fermi theory, in which the lepton pair is created only via the V interaction.

- After having used (2.43) in the lepton trace, the remaining term in γ_5 produces an antisymmetric tensor in μ and ν, which gives zero when the indices μ and ν are contracted with the corresponding indices of the hadronic currents in equation (2.42).

- As in neutron decay, we can approximate the final momentum of the pion in the non-relativistic limit, setting:

$$\frac{(p_i + p_f)_\mu}{2\sqrt{E_i E_f}} = g_{\mu 0}. \tag{2.44}$$

We therefore obtain:

$$d\Gamma = (2\pi)^4 \delta^{(4)}(P_{out} - P_{in}) \left(\frac{G_F}{\sqrt{2}}\right)^2| < I^{(+)} > |^2 \times$$

$$2 \left(\frac{1}{E_e E_\nu}\right) [2 E_e E_\nu - (e\nu)] \frac{d^3 p_f}{(2\pi)^3} \frac{d^3 p_e}{(2\pi)^3} \frac{d^3 p_\nu}{(2\pi)^3}. \tag{2.45}$$

The calculations are easily carried out in the non-relativistic limit for the final pion:

- In the π^- rest system, the function $\delta^{(3)}(\sum \mathbf{P}_{out})$ is integrated over the π^0 momentum.

- Conservation of energy does not depend on the angles and can be expressed:

$$E_e + E_\nu = Q \equiv m(\pi^+) - m(\pi^0) = 4.59 \text{ MeV}. \tag{2.46}$$

- The integrations over the angles of the electron and neutrino can be done independently; note that $\int (e \cdot \nu) d\cos\theta_\nu = 2E_e E_\nu$.

- Integration over the neutrino energy makes use of the δ-function for conservation of energy and sets $E_\nu = Q - E_e$.

The final result is

$$\Gamma = \frac{G_F^2}{\pi^3} \int_{m_e}^{Q} dE \sqrt{E^2 - m_e^2} E(Q - E)^2 \simeq (\frac{G_F^2 Q^5}{30\pi^3})0.941 \simeq 0.41 \text{ s}^{-1}. \tag{2.47}$$

where we have used:

$$G_F^{(n)} = 0.9740 \, G^\mu = 1.136 \, 10^{-5} \text{ GeV}^{-2}. \tag{2.48}$$

Some small electromagnetic corrections should be added to the result (2.47), but it already compares very well with the experimental value:

$$\Gamma_{expt} = B(\pi^+ \to \pi^0 + e^+ + \nu_e)/\tau_\pi = 0.39 \text{ s}^{-1}. \tag{2.49}$$

2.5 $SU(3)$ SYMMETRY

$SU(3)$ is the group of U matrices which are unitary and 3×3; it was considered for the first time by Sakata to generalise the isotopic spin transformation of the nucleon doublet to include strangeness.

Adding the Λ baryon to the nucleons, Sakata considered the *triplet* of fields:

$$\begin{pmatrix} p \\ n \\ \Lambda \end{pmatrix} \tag{2.50}$$

and extended (2.2) to the transformations of the triplet:

$$\begin{pmatrix} p \\ n \\ \Lambda \end{pmatrix} \to U \begin{pmatrix} p \\ n \\ \Lambda \end{pmatrix} \tag{2.51}$$

with U unitary, 3×3 and $det(U) = 1$.

The symmetry under $SU(3)$ is reduced to $SU(2) \otimes U(1)_S$ by the mass difference between the Λ baryon ($M_\Lambda = 1115.7$ MeV) and the mass of the nucleons, an effect of around 30%; the $SU(3)$ symmetry is broken by a component of the strong interaction itself, often referred to as the *medium strong interaction.* Consequently, we should expect that the relations derived from the symmetry to be much less precise than those related to isotopic spin.

In the Standard Theory the fundamental triplet is that of the up, down and strange quarks, and the breaking of the $SU(3)$ symmetry is a result of the mass difference between the strange quark and the average mass of the up and down quarks.

The Gell–Mann matrices.   Every unitary matrix can be written as an exponential of a Hermitian matrix:

$$U = e^{iT}. \tag{2.52}$$

If in addition U has determinant equal to unity, it must be true that:

$$Tr(T) = 0. \tag{2.53}$$

We can expand every 3×3 Hermitian matrix in a basis of standard Hermitian matrices, in analogy with the expansion of a Hermitian 2×2 matrix in the basis of the Pauli matrices plus the identity matrix.

In the case of $SU(3)$, there are 8 traceless Hermitian matrices. A convenient choice of basis has been identified by Gell–Mann in the matrices λ_i, $(i = 1, \cdots, 8)$:

- i = 1, 2, 3 (τ_i are the three 2×2 Pauli matrices):

$$\lambda_i = \begin{pmatrix} \tau_i & 0 \\ 0 & 0 \end{pmatrix}, \tag{2.54}$$

- i = 4, 5:

$$\lambda_4 = \begin{pmatrix} 0 & 0 & 1 \\ 0 & 0 & 0 \\ 1 & 0 & 0 \end{pmatrix}; \lambda_5 = \begin{pmatrix} 0 & 0 & -i \\ 0 & 0 & 0 \\ i & 0 & 0 \end{pmatrix}, \tag{2.55}$$

- i = 6, 7:

$$\lambda_6 = \begin{pmatrix} 0 & 0 & 0 \\ 0 & 0 & 1 \\ 0 & 1 & 0 \end{pmatrix}; \lambda_7 = \begin{pmatrix} 0 & 0 & 0 \\ 0 & 0 & -i \\ 0 & i & 0 \end{pmatrix}, \tag{2.56}$$

- i = 8

$$\lambda_8 = \frac{1}{\sqrt{3}} \begin{pmatrix} 1 & 0 & 0 \\ 0 & 1 & 0 \\ 0 & 0 & -2 \end{pmatrix}. \tag{2.57}$$

The Gell–Mann matrices are normalised like the Pauli matrices:

$$\text{Tr}(\lambda_i \lambda_j) = 2\delta_{ij} \ . \tag{2.58}$$

In the set of Gell–Mann matrices, there are two of them which commute with each other, which are λ_3 and λ_8. This indicates that $SU(3)$ is a group of rank two: two generators can be simultaneously diagonalised. $SU(3)$ is a good candidate to describe a symmetry which includes isotopic spin and strangeness as conserved quantum numbers.

The generator associated with λ_3 corresponds to the third component of isotopic spin:

$$I_3 = \frac{1}{2}\lambda_3 = \begin{pmatrix} \frac{1}{2} & 0 & 0 \\ 0 & -\frac{1}{2} & 0 \\ 0 & 0 & 0 \end{pmatrix} \ . \tag{2.59}$$

The generator commuting with the total isotopic spin is clearly λ_8, which we connect to hypercharge according to:

$$Y = \frac{1}{\sqrt{3}}\lambda_8 = \begin{pmatrix} \frac{1}{3} & 0 & 0 \\ 0 & \frac{1}{3} & 0 \\ 0 & 0 & -\frac{2}{3} \end{pmatrix} \ . \tag{2.60}$$

If we use equation (2.16) we also obtain:

$$Q = \begin{pmatrix} \frac{2}{3} & 0 & 0 \\ 0 & -\frac{1}{3} & 0 \\ 0 & 0 & -\frac{1}{3} \end{pmatrix} \ . \tag{2.61}$$

The physical interpretation of the numerical coefficients in (2.60) and (2.61) will be clarified in chapter 8.

YANG–MILLS THEORY

CONTENTS

Electrodynamics is invariant under so-called gauge transformations of the second type:

$$\psi(x) \to \psi'(x) = e^{iq\phi(x)}\psi(x); \quad A_\mu(x) \to A'_\mu(x) = A_\mu(x) - \partial_\mu\phi(x) \tag{3.1}$$

where ϕ is an arbitrary function of space-time, ψ is a generic matter field, q is its electric charge in units of the elementary charge (for the electron field, $q = -1$) and A_μ is proportional to the electromagnetic field.

The derivatives $\partial_\mu\psi$ do not transform like ψ, therefore, while the free Lagrangian of ψ is invariant under *global* phase transformations $\phi(x) =$ constant, the same Lagrangian is not invariant for the transformations (3.1).

However, we can define a *covariant derivative* through the minimal substitution:

$$\partial_\mu\psi(x) \to D_\mu\psi(x) = [\partial_\mu + iqA_\mu(x)]\psi(x) \tag{3.2}$$

and then the field $D_\mu\psi$ transforms exactly like ψ:

$$(D_\mu\psi)' = [\partial_\mu + iqA_\mu(x) - iq\partial_\mu\phi(x)][e^{iq\phi(x)}\psi(x)] =$$
$$= e^{iq\phi(x)}[\partial_\mu + iqA_\mu(x)]\psi(x)$$
$$= e^{iq\phi(x)}D_\mu\psi(x). \tag{3.3}$$

A_μ also transforms in a complicated way but the Maxwell tensor $F_{\mu\nu} = \partial_\nu A_\mu - \partial_\mu A_\nu$, which contains the observable electric and magnetic fields, is invariant:

$$(F_{\mu\nu})' = F_{\mu\nu} - \partial_\nu\partial_\mu\phi + \partial_\mu\partial_\nu\phi = F_{\mu\nu}. \tag{3.4}$$

In conclusion:

- If the matter Lagrangian $\mathcal{L}_0(\psi, \partial_\mu\psi)$, for example the free Lagrangian, is invariant for global phase transformations, the new Lagrangian:

$$\mathcal{L} = \mathcal{L}_0(\psi, D_\mu\psi) \tag{3.5}$$

is invariant for transformations of the second kind, (3.1).

- A complete dynamical theory of the matter field ψ in interactions with the electromagnetic field is obtained by adding to (3.5) the invariant Lagrangian obtained from the Maxwell tensor:

$$\mathcal{L}_{min} = \mathcal{L}_0(\psi, D_\mu\psi) - \frac{1}{4e^2}F_{\mu\nu}F^{\mu\nu}. \tag{3.6}$$

- e is a constant, the only quantity in (3.6), apart from the mass of the particles associated with ψ, which is not determined by the symmetry. Rescaling $A \to eA$, A simply becomes the vector potential of the electromagnetic field, while e characterises the strength of the interaction between ψ and the electromagnetic field.

The minimal substitution applied to the Dirac Lagrangian of the charged leptons, electron, μ and τ, gives us so-called spinor electrodynamics (or QED [1]):

$$\mathcal{L} = \mathcal{L}_0(\psi_e, D_\mu\psi_e) + (e \to \mu) + (e \to \tau)$$
$$\mathcal{L}_0(\psi, D_\mu\psi) = \bar{\psi}(i\partial_\mu\gamma^\mu - m)\psi + e\bar{\psi}A_\mu\gamma^\mu\psi. \tag{3.7}$$

The local gauge symmetry, (3.1), completely determines the form of the interaction (for example it predicts for the gyromagnetic ratio of the electron the value from the Dirac theory $g = 2$).

Note. A detailed account of the origin of gauge theories can be found in [6]. In brief, we recall that gauge transformations were introduced in 1918 by Weyl [7], in the context of the theory of general relativity, as transformations dependent on the local coordinate scale:

$$dx \to e^{\lambda(x)}dx, \qquad \lambda(x) = \int^x A_\mu(y)dy^\mu \tag{3.8}$$

where λ are real functions. The laws of physics, however, are not invariant under scale transformations, as immediately noted by Einstein, and the theory was abandoned. After the arrival of quantum mechanics, London, in his formulation of the theory of superconductivity, observed that the minimal substitution of classical electromagnetism [1] gave rise to the transformation (3.2) which we can interpret as the effect of the transformation:

$$\psi(x) \to e^{i\lambda(x)}\psi(x), \qquad \lambda(x) = \int^x A_\mu(y)dy^\mu \tag{3.9}$$

on the wave function of the electron. Weyl, in 1929, accepted the crucial introduction of the imaginary unit and then proposed that the substitution (3.1) should be used as the principle from which to derive the laws of electrodynamics, to which the name *gauge principle* or *minimal principle* was given. The work of Weyl in 1929 was the beginning of modern gauge theories. The idea that symmetry could generate dynamics was taken up by Yang, then a student in Chicago, after reading an article by Pauli which explained Weyl's work.

3.1 NON-COMMUTATIVE LOCAL SYMMETRIES

We consider a field ψ which transforms under a non-abelian group G according to the equation:

$$\psi(x) \to U\psi(x) \tag{3.10}$$

with U constant, independent of the location in space-time.

Following Yang and Mills [8], we can argue that a global symmetry like (3.10) is highly unnatural in a local field theory. Symmetry tells us that the choice of the components of ψ in the space of the internal quantum numbers (e.g. the distinction between proton and neutron) is purely conventional; the strong interactions should not depend on which linear combination of p and n we choose to define the *physical proton*. For a global symmetry, however, we must fix that convention in the whole of space-time and we can change it only in strict agreement with all observers distributed in different regions of the universe. But, in a local field theory, each observer is influenced only by the fields which are in their immediate vicinity *(here and now)* and should not need to be aware of the choices made by distant observers, outside of that observer's own past light cone.

In this way of looking at things, it seems natural to require that the symmetry under G should be a *local* symmetry; the physics should be invariant for transformations:

$$\psi(x) \to U(x)\psi(x) \tag{3.11}$$

where $U(x)$ is an element of G *variable in an arbitrary way* with the location in space-time.

The problem to be solved, for a symmetry of the type (3.11) is the same as in electrodynamics: how to treat the derivatives of ψ, or how to compare fields at different points of space-time, given that each one of them can be changed, according to (3.11), for group transformations arbitrarily different from one another. The solution repeats what was found for electrodynamics: the construction of a covariant derivative.

To arrive at this result we must introduce the notion of *parallel transport*[1].

Starting from $\psi(x)$ we want to define a new field, $\psi(x, dx)_{TP}$, which should be the field $\psi(x)$ translated to $x + dx$ in a *parallel fashion*, in the sense that it transforms as would $\psi(x + dx)$:

$$\psi(x, dx)_{TP} \text{ is such that } \psi(x, dx)'_{TP} = U(x + dx)\psi(x, dx)_{TP} . \tag{3.12}$$

$\psi(x, dx)_{TP}$ must be linear in $\psi(x)$ and agree with itself for $dx = 0$. To first order in dx we can write:

$$\psi(x, dx)_{TP} = (1 - i\Gamma_\mu(x)dx^\mu)\psi(x) . \tag{3.13}$$

---

[1] This argument follows the reasoning made in the theory of general relativity, in which a similar problem arises concerning the components of non-invariant quantities, for example 4-vectors.

We now require (3.12) to hold. The right hand side is equal to:

$$U(x+dx)\psi(x,dx)_{TP} = U(x)\psi(x) - iU(x)\Gamma_\mu dx^\mu + dU(x)\psi(x) \qquad (3.14)$$

which must be equal to:

$$\psi(x,dx)'_{TP} = (1 - i\Gamma'_\mu dx^\mu)U(x)\psi(x) = U(x)\psi(x) - i\Gamma'_\mu U(x)dx^\mu \psi(x). \quad (3.15)$$

Setting the first order terms to be equal, therefore gives:

$$\Gamma'_\mu = U(x)\Gamma_\mu U^{-1}(x) + i[\partial_\mu U(x)]U^{-1}(x). \qquad (3.16)$$

Once $\psi(x,dx)_{TP}$ has been constructed, we can form the differential:

$$\psi(x+dx) - \psi(x,dx)_{TP} = (\partial + i\Gamma_\mu)\psi(x)dx^\mu = D_\mu\psi(x)dx^\mu \qquad (3.17)$$

which clearly transforms like $\psi(x)$ for gauge transformations:

$$D_\mu\psi(x) \to D'_\mu\psi(x)' = U(x)D_\mu\psi(x). \qquad (3.18)$$

Once in possession of the covariant derivative, we can promote a Lagrangian invariant under global transformations to one invariant under local transformations by means of the *minimal substitution*, exactly as in electrodynamics:

$$\mathcal{L}_0(\psi, \partial_\mu\psi) \to \mathcal{L}_0(\psi, D_\mu\psi) \qquad (3.19)$$

The matrices $\Gamma_\mu(x)$, the *connections* of the field between x and $x + dx$, in their turn define fields analogous to the electromagnetic field in (3.1), the *gauge fields*.

For a pure phase transformation (corresponding to an abelian group G):

$$U = e^{iq\phi}; \qquad i[\partial_\mu U(x)]U^{-1}(x) = -q\partial_\mu\phi(x)$$
$$\Gamma_\mu = qA_\mu \, . \qquad (3.20)$$

In general, we set:

$$\Gamma_\mu(x) = A^i_\mu T^i; \qquad (3.21)$$
$$U = 1 + i\alpha_i T^i \qquad (3.22)$$

with T^i the generators of the infinitesimal transformation of G and α_i the parameters of that transformation (with the sum over repeated indices understood). In the infinitesimal case, (3.16) is written:

$$(A^i_\mu)'T^i = A^i_\mu T^i - iA^i_\mu \alpha_j[T^i, T^j] - \partial_\mu\alpha_i T^i \qquad (3.23)$$

and, taking account of the commutation rules of the group:

$$[T^i, T^j] = if^{ijk}T^k \, , \qquad (3.24)$$

we finally find:

$$(A^i_\mu)' = A^i_\mu - f^{ils}\alpha_l A^s_\mu - \partial_\mu \alpha_i \ . \tag{3.25}$$

Under global transformations, $\alpha_i = constant$, we see from (3.25) that the connections transform like the *regular (or adjoint) representation* of the group (cf. Appendix C):

$$(A^i_\mu)' = A^i_\mu - f^{ils}\alpha_l A^s_\mu = A^i_\mu - i\alpha_l(-if^{ils})A^s_\mu =$$
$$= A^i_\mu - i\alpha_l(K^l_{reg})_{is} A^s_\mu; \qquad (\alpha_i = \text{constant}). \tag{3.26}$$

Note. The result (3.25) tells us that the transformation law of the gauge fields is *universal*, i.e. independent of the specific representation of the group G realised by the matrices U on the fields ψ. In effect, in the presence of different matter fields, we should consider the case in which the transformation law (3.11) is realised differently on different fields:

$$\psi^{(A)}(x) \to (\psi^{(A)}(x))' = U^{(A)}(g(x))\psi^{(A)}(x); \qquad A = 1, 2, \ldots, N; \tag{3.27}$$

where the $U^{(A)}(g)$ are irreducible representations of the group G. In this case the local symmetry transformation is identified by an element of the group[2], $g(x)$, and realised by the representation $U^{(A)}(g)$ on the fields $\psi^{(A)}$.

The Yang–Mills tensor. To construct a Lagrangian invariant for the gauge fields, we must use the fields with definite transformation properties. Starting from the Γ_μ connections we introduce[3] the *Yang–Mills tensor*:

$$G_{\mu\nu} = \partial_\nu \Gamma_\mu - \partial_\mu \Gamma_\nu + i[\Gamma_\nu, \Gamma_\mu] \ . \tag{3.28}$$

Similarly to what was done in (3.21), we introduce the fields $G^i_{\mu\nu}$

$$G_{\mu\nu} = G^i_{\mu\nu} T^i$$
$$G^i_{\mu\nu} = \partial_\nu A^i_\mu - \partial_\mu A^i_\nu + f^{ijk} A^j_\mu A^k_\nu \ . \tag{3.29}$$

It is not difficult to show that:

$$G'_{\mu\nu} = \partial_\nu A'_\mu - \partial_\mu A'_\nu + i[A'_\nu, A'_\mu] = U G_{\mu\nu} U^{-1} \tag{3.30}$$

and that therefore the Yang–Mills Lagrangian

$$\mathcal{L}_{Y-M} = -\frac{1}{8g^2} Tr(G_{\mu\nu} G^{\mu\nu}) \tag{3.31}$$

is invariant.

---

[2] In the language of the previous section we can consider that the transformation g acts on the fundamental degrees of freedom of the Lagrangian, while the $U^{(A)}(g)$ represent the action of the same transformation on the fields N, π, etc.

[3] Note that the order in which μ and ν appear on the two sides is opposite, as in QED.

As for the gauge fields, the properties of the Yang–Mills tensor are independent of the representation chosen for the matrices U. Limiting ourselves to infinitesimal transformations, equation (3.30) leads to:

$$(G^i_{\mu\nu})' = G^i_{\mu\nu} - f^{ils}\alpha_l G^s_{\mu\nu} . \tag{3.32}$$

The Yang–Mills tensor *transforms like the regular representation*, even for local transformations.

With the normalisation:

$$Tr(T^i T^j) = 2\delta^{ij}. \tag{3.33}$$

the Lagrangian (3.31) takes the form:

$$\mathcal{L}_{Y-M} = -\frac{1}{8g^2}G^i_{\mu\nu}(G^j)^{\mu\nu}Tr(T^iT^j) =$$

$$= -\frac{1}{4g^2}G^i_{\mu\nu}(G^i)^{\mu\nu}. \tag{3.34}$$

3.2   MINIMAL LAGRANGIAN

The extension of a global symmetry to local non-abelian gauge transformations finally gives the minimal Lagrangian.

$$\mathcal{L} = \mathcal{L}_0(\psi, D_\mu\psi) - \frac{1}{4g^2}G^i_{\mu\nu}(G^i)^{\mu\nu} \tag{3.35}$$

The analogy with QED is clear, but there are also important differences which we describe in what follows.

If the group G is a *simple group*, i.e. not factorisable into mutually commuting subgroups of transformations, the constant g is the only arbitrary constant, other than any possible masses contained in $\mathcal{L}_0$. In the opposite case, the group is called *semi-simple* and there are as many independent constants as there are commuting factors of the group. For example, the group $SU(2)_I \otimes U(1)_Y$ would have two constants, g and g'.

Rescaling $A \to gA$, we see that g is a coupling constant. We give the forms of the various rescaled terms:

$$D_\mu\psi = (\partial_\mu + igA^i_\mu T^i)\psi \qquad \text{(covariant derivative);} \tag{3.36}$$

$$G^i_{\mu\nu} = \partial_\nu A^i_\mu - \partial_\mu A^i_\nu + gf^{ijk}A^j_\mu A^k_\nu \quad \text{(Yang − Mills tensor).} \tag{3.37}$$

Correspondingly, the Yang–Mills action is written explicitly as:

$$\mathcal{L}_{Y-M} = -\frac{1}{4}G^i_{\mu\nu}(G^i)^{\mu\nu} =$$

$$= \frac{1}{2}A^i_\mu(g^{\mu\nu}\Box - \partial^\mu\partial^\nu)A^i_\nu$$

$$+ g\, f^{ijk}[\partial^\nu(A^i)^\mu]A^j_\nu A^k_\mu + g^2 f^{ilm}f^{isk}A^l_\nu A^m_\mu(A^s)^\nu(A^s)^\mu \tag{3.38}$$

while, for the Dirac and Klein–Gordon Lagrangians, we have:

$$\mathcal{L}_D = \bar{\psi}(i\not{D} - M)\psi \;=\; \bar{\psi}(i\not{\partial} - M)\psi - gA^i_\mu\,\bar{\psi}\gamma^\mu T^i\psi\,, \qquad (3.39)$$

$$\begin{aligned}\mathcal{L}_{KG} &= \phi^\dagger(-D_\mu D^\nu - \mu^2)\phi = \\ &= \phi^\dagger(-\Box - \mu^2)\phi - gA^i_\mu(\phi^\dagger T^i i\partial^\mu\phi - i\partial^\mu\phi^\dagger T^i\phi) \\ &\quad + \frac{1}{2}g^2 A^i_\mu(A^j)^\mu\phi^\dagger\{T^i, T^j\}\phi\,.\end{aligned} \qquad (3.40)$$

The notable fact is that the form of the interaction between matter and gauge fields is fixed by the symmetry.

The gauge fields interact among themselves (i.e. they change in a non-trivial way under global transformations). This is an important difference compared to QED, which makes the Yang–Mills theory rather similar to gravity. Even in the absence of matter, the Yang–Mills theory is non-trivial, actually a simplified version of Einstein's theory of gravity.

In the limit $g = 0$, the Yang–Mills Lagrangian (3.34) and (3.38) describes free particles, as many as there are generators of the group, each one characterised by a Lagrangian identical to the Maxwell Lagrangian of the free electromagnetic field:

- *For $g = 0$, the quanta of the gauge field are particles of zero mass and spin 1.*

Equations of motion. These are the Euler–Lagrange equations derived starting from equation (3.35). Assuming a Dirac Lagrangian for $\mathcal{L}_0$, we obtain the derivatives:

$$\frac{\partial\mathcal{L}}{\partial\partial_\mu A^i_\alpha} = -G^i_{\alpha\mu} \qquad (3.41)$$

$$\frac{\partial\mathcal{L}}{\partial A^i_\alpha} = -\bar{\psi}\gamma_\alpha T^i\psi \;-\; g\,f^{ijk}(A^j)^\sigma\,G^k_{\alpha\sigma} \qquad (3.42)$$

and equations of motion:

$$\begin{aligned}[iD_\mu\gamma^\mu + m]\psi &= 0 &&\text{(fermion field);} &&(3.43)\\ \partial^\mu G^i_{\alpha\mu} &= \bar{\psi}\gamma_\alpha T^i\psi \;+\; g\,f^{ijk}(A^j)^\sigma\,G^k_{\alpha\sigma} &&\text{(gauge fields).} &&(3.44)\end{aligned}$$

Taking the terms which contain the gauge fields to the left hand side of (3.44) reconstructs the covariant derivative of the Yang–Mills tensor:

$$(D_\rho)^{ik}G^k_{\mu\nu} = (\partial_\rho\delta^{ij} - gfijkA^j_\rho)G^k_{\mu\nu} \qquad (3.45)$$

and we obtain the explicit covariant form of (3.44) (what else could we have expected?):

$$(D^\mu\,G_{\alpha\mu})^i = \bar{\psi}\gamma_\alpha T^i\psi. \qquad (3.46)$$

Equation (3.46) does *not* lead to a continuity equation for the matter current, because the covariant derivatives *do not commute with each other.* To obtain a conserved current we must return to (3.44). Deriving equation (3.44), indeed we find, as a result of the antisymmetry of $G_{\alpha\mu}$, a continuity equation for the current:

$$\mathcal{J}_\mu^i = \bar\psi \gamma_\alpha T^i \psi + g\, f^{ijk} (A^j)^\sigma\, G_{\alpha\sigma}^k, \tag{3.47}$$

$$\partial^\alpha \partial^\mu G_{\alpha\mu}^i = \partial^\alpha \mathcal{J}_\alpha^i = 0. \tag{3.48}$$

It is easily recognised that $\mathcal{J}$ is the total Noether current, including the contribution of the gauge fields. Only the total current is conserved, because charge can flow from matter to the gauge fields. This result is analogous to what happens to the energy-momentum tensor in the theory of general relativity; only the energy and momentum of the overall matter plus gravitational field are conserved.

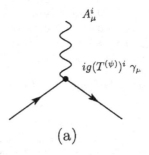

(a)

Figure 3.1 Spin $\frac{1}{2}$ vertex.

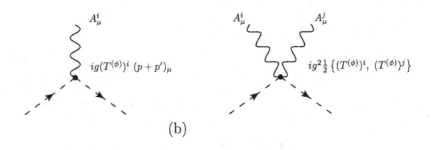

(b)

Figure 3.2 Spin 0 vertices.

Vertices. We consider matter fields of spin $\frac{1}{2}$ and spin 0 (for example the nucleon and the pion) in interactions with gauge fields. In a perturbative treatment, the part of the interaction in (3.34) and (3.38) determines the *vertices* of the corresponding Feynman diagrams [3]. From (3.38) we see that we have vertices of first order and, for scalar fields, second order in g. The first order vertices are determined by the Noether matter current:

$$\mathcal{L}_{int}^{(1)} = gA_\mu^i(J^i)^\mu \ ;$$
$$J_\mu^i = \bar{\psi}\gamma_\mu(T^{(N)})^i\psi \ + \ i[\phi^\dagger(T^{(\pi)})^i\partial_\mu\phi - \partial_\mu\phi^\dagger(T^{(\pi)})^i\phi] \ . \tag{3.49}$$

The diagrams corresponding to the two types of vertex are shown in Figures 3.1 and 3.2. For the spin 0 fields, in addition to the interaction mediated by the Noether current, the interaction Lagrangian contains a term quadratic in the gauge fields[4]. The relevant diagram is shown in Figure 3.2.

Problem 1. Prove equation (3.30).

Problem 2. Starting from the Lagrangian (3.35), derive the current (3.47) from Noether's theorem relevant to the global transformations (3.10) and (3.26).

Problem 3. Show that in an invariant expression the covariant derivatives can also be *integrated by parts*

$$(D_\mu\phi)^\dagger\ldots\phi = -\phi^\dagger D_\mu(\ldots\phi) + \partial_\mu(\phi^\dagger\ldots\phi). \tag{3.50}$$

---

[4]The corresponding diagram is commonly known as a *seagull diagram*, from its shape which resembles a gull in flight.

ELECTROWEAK UNIFICATION I

CONTENTS

After publication in 1954, the Yang–Mills theory remained for a long time an idea in search of an application.

The success of isotopic spin symmetry had led to a belief that the theory of strong interactions would be the natural field of application. The spin 1 mesons of Figure 2.2 had been identified as the gauge fields of the strong interaction. However, the theory of *vector meson dominance* had never gone beyond some phenomenological success and limited predictive power.

A second line of thought, initiated by Schwinger, pointed to the weak and electromagnetic interactions as the natural field of applications for the ideas of Yang and Mills.

The success of V–A theory reinforced the idea that the weak interactions of Fermi were mediated by vector (gauge) fields, the *intermediate vector bosons* (cf. [1]). Moreover, according to the CVC hypothesis of Feynman and Gell–Mann, the vector bosons should be coupled, at least for the V part, to the Noether current of isotopic spin *just as required by a gauge theory*.

It is also possible to hypothesise the unification of the weak and electromagnetic interactions in a Yang–Mills theory which includes the photon and intermediate bosons in the same symmetry. Isotopic spin symmetry, so prominent in nuclear phenomena, would then be the foundation of a gauge theory of the weak and electromagnetic interactions!

The main obstacle to this line is represented by the mass of the intermediate bosons. As we saw, the theory predicts that the gauge fields have zero mass like the photon, at least in the $g = 0$ limit. Conversely, the masses of the intermediate bosons must be large enough that these particles do not cause visible effects in the weak decays of the neutron and pion, etc.[1]. How to *give a mass* to the bosons in a theory with weak coupling ($g \simeq e$) remained a mystery for a long time. The problem was resolved in a satisfactory way only around the middle of the 1960s, with the discovery of the so-called *Brout–Englert–Higgs mechanism*, linked to spontaneous breaking of the gauge symmetry.

A provisional but effective solution was to insert into the Lagrangian *ad hoc* mass terms for the vector bosons, assuming that the gauge symmetry could be explicitly violated by these masses (as isotopic spin symmetry is violated by the proton-neutron mass difference) without losing its main properties. Even if this hope is revealed unfounded (the theory proved a posteriori to be non-renormalisable and therefore mathematically inconsistent) the Yang–Mills theory with ad hoc masses was an important tool to explore the phenomenological properties of electroweak unification. The theory of Glashow in 1961 [10] identified for the first time the appropriate gauge group to describe the electroweak interactions, the group $SU(2) \otimes U(1)$, with the consequent necessity of a neutral intermediate boson, the Z^0, in addition to the charged bosons, $W^{\pm}$, and the photon.

In this chapter we limit ourselves to the interactions of leptons and we describe the Glashow theory for essentially pedagogic reasons. We delay to a later chapter the discussion of the theory proposed by Weinberg and Salam [13] in which the masses are generated in a satisfactory way by the spontaneous breaking of gauge symmetry.

4.1 SYMMETRIES OF THE ELECTRON DOUBLET

In this section we consider the electron neutrino-electron doublet. We would like to identify the symmetries of this system, that (i) commute with the Lorentz group (i.e. are *internal* symmetries), and (ii) contain, within the current determined by Noether's theorem, the weak current and the electromagnetic current[1]:

$$
\begin{aligned}
J_\mu^W &= \bar{e}\gamma_\mu(1 - \gamma_5)\nu_e \; ; \\
J_\mu^{e.m.} &= -\bar{e}\gamma_\mu e \; .
\end{aligned}
\tag{4.1}
$$

By analogy with the isotopic spin formalism, we collect the fields ν_e and e in a doublet:

$$
\psi = \begin{pmatrix} \nu_e \\ e \end{pmatrix} .
$$

The transformations that operate on the components of the doublet are written:

$$
\psi \to e^{i\alpha \cdot \frac{\tau}{2}}\psi \; .
\tag{4.2}
$$

Regarding the Dirac space, the transformations (4.2) contain only the identity matrix which obviously commutes with the generators of the Lorentz

---

[1]The definition of the electromagnetic current corresponds to assigning to the electron charge -1, in units of the fundamental charge $e > 0$, equal to the electric charge of the proton.

group, the matrices $\sigma^{\mu\nu}$. Therefore equation (4.2) defines an internal symmetry.

However, there is another Dirac matrix, γ_5, which is still invariant if we restrict ourselves to *proper* Lorentz transformations[2]. We can therefore define a second set of internal transformations of the doublet, the *chiral* transformations:

$$\psi \rightarrow e^{i\beta \cdot \frac{\tau}{2}\gamma_5}\psi .$$

For infinitesimal transformations.

$$\psi' = \psi + i\alpha \cdot \frac{\tau}{2}\psi ;$$

$$\psi' = \psi + i\beta \cdot \frac{\tau}{2}\gamma_5\psi . \tag{4.3}$$

To simplify the notation, we introduce the projectors of the chiral fields (we recall that $\gamma_5^2 = 1$):

$$a^{(\pm)} = \frac{1 \pm \gamma_5}{2};$$

$$(a^{(\pm)})^2 = a^{(\pm)}; \quad a^{(+)}a^{(-)} = 0 \tag{4.4}$$

and denote with $\psi_{L,R}$ the corresponding chiral fields[3]:

$$\psi_L = a^{(-)}\psi ; \qquad \psi_R = a^{(+)}\psi . \tag{4.5}$$

Finally we collect together the generators of the transformations (4.2) and (4.3) by constructing the so-called *chiral generators*:

$$(\mathcal{I}^{(\pm)})^i = a^{(\pm)}\frac{\tau^i}{2} . \tag{4.6}$$

The commutation relations are immediately found:

$$[(\mathcal{I}^{(\pm)})^i, (\mathcal{I}^{(\pm)})^j] = i\epsilon^{ijk}(\mathcal{I}^{(\pm)})^i;$$

$$[(\mathcal{I}^{(+)})^i, (\mathcal{I}^{(-)})^j] = 0 . \tag{4.7}$$

The transformations (4.3) characterise an algebra $SU(2)_L \otimes SU(2)_R$ (the indices R and L indicate that the first $SU(2)$ acts only on the fields ψ_L and the second on the fields ψ_R).

---

[2]The proper and isochronous Lorentz transformations (cf. [1]) are the true group of space-time symmetry, given that parity and time reversal are not conserved in the weak interactions; therefore the fact that the γ_5 matrix transforms under parity like a pseudoscalar does not exclude it.

[3]The indices L, R stand respectively for left-handed and right-handed. In the limit of zero mass, the field ψ_L destroys a fermion with negative helicity (the spin is equivalent to an anticlockwise rotation around the direction of motion) and creates an antifermion with positive helicity, while ψ_R destroys a fermion with positive helicity (clockwise rotation) and creates an antifermion with negative helicity (cf. [1]).

The representations of $SU(2)_L \otimes SU(2)_R$ are characterised by two integer or half-integer quantum numbers, I_1 and I_2, the isospin associated with each of the two $SU(2)$ factors. In particular, for ψ_L and ψ_R we have:

$$\psi_L \simeq (1/2, 0); \qquad \psi_R \simeq (0, 1/2). \tag{4.8}$$

To the transformations (4.3) we can add abelian transformations by a constant phase factor (of ν_e and e) multiplied by unity or by γ_5. In terms of infinitesimal transformations:

$$\psi' = \psi + i\alpha_0 \psi; \qquad \psi' = \psi + i\beta_0 \gamma_5 \psi \tag{4.9}$$

The transformations (4.9) add two abelian factors to the symmetry group, the first of which corresponds to the conservation of electron number $L_e = N(e) - N(\bar{e}) + N(\nu_e) - N(\bar{\nu}_e)$ (cf. [1]).

In conclusion, the transformations that we can carry out on the doublet of Dirac fields (ν_e, e) form an algebra: $SU(2)_L \otimes SU(2)_R \otimes U(1)_{L_e} \otimes U(1)_A$.

4.2 MINIMAL GAUGE GROUP

The Noether currents associated with the transformations of $SU(2)_L \otimes SU(2)_R \otimes U(1)_{l_e} \otimes U(1)_A$ are obtained from the free Lagrangian of e and ν_e. Assuming a massless neutrino, we write[4]:

$$\begin{aligned}
\mathcal{L}_0 &= \mathcal{L}_{00} + \mathcal{L}_m \ ; \\
\mathcal{L}_{00} &= \bar{e}i\slashed{\partial}e + \bar{\nu}i\slashed{\partial}\nu = \bar{e}_L i\slashed{\partial}e_L + \bar{\nu}_L i\slashed{\partial}\nu_L + \bar{e}_R i\slashed{\partial}e_R + \bar{\nu}_R i\slashed{\partial}\nu_R = \\
&= \bar{\psi}_L i\slashed{\partial}\psi_L + \bar{\psi}_R i\slashed{\partial}\psi_R \ ; \\
\mathcal{L}_m &= m_e \bar{e}e = m_e(\bar{e}_L e_R + \bar{e}_R e_L) \ .
\end{aligned} \tag{4.10}$$

and we find the eight currents:

$$\mathbf{L}_\mu = -i\frac{\partial \mathcal{L}_{00}}{\partial \partial_\mu \psi}a^{(-)}\frac{\tau}{2}\psi = \bar{\psi}_L \gamma_\mu \frac{\tau}{2}\psi_L; \quad L^0_\mu = -i\frac{\partial \mathcal{L}_{00}}{\partial \partial_\mu \psi}a^{(-)}\psi = \bar{\psi}_L \gamma_\mu \psi_L$$

$$\mathbf{R}_\mu = -i\frac{\partial \mathcal{L}_{00}}{\partial \partial_\mu \psi}a^{(+)}\frac{\tau}{2}\psi = \bar{\psi}_R \gamma_\mu \frac{\tau}{2}\psi_R; \quad R^0_\mu = -i\frac{\partial \mathcal{L}_{00}}{\partial \partial_\mu \psi}a^{(+)}\psi = \bar{\psi}_R \gamma_\mu \psi_R.$$

$$\tag{4.11}$$

The form of the weak current in (4.1) suggests to identify it with the current corresponding to the raising operator of $SU(2)_L$:

$$L^1_\mu + iL^2_\mu = \bar{e}_L \gamma_\mu \nu_L = \frac{1}{2}\bar{e}\gamma_\mu(1 - \gamma_5)\nu \ . \tag{4.12}$$

---

[4]In this section, for brevity, we omit the index e from the neutrino field and we recall that $\psi = \psi_L + \psi_R$, $\bar{\psi}_L \gamma_\mu \psi_R = \bar{\psi}_R \gamma_\mu \psi_L = 0$.

However, the neutral current of $SU(2)_L$, L^3_μ, does not coincide with the electromagnetic current. Instead we have:

$$J^{e.m.}_\mu - L^3_\mu = -(\bar{e}_L\gamma_\mu e_L + \bar{e}_R\gamma_\mu e_R) - \frac{1}{2}(\bar{\nu}_L\gamma_\mu\nu_L - \bar{e}_L\gamma_\mu e_L) =$$

$$= -\bar{e}_R\gamma_\mu e_R - \frac{1}{2}(\bar{\nu}_L\gamma_\mu\nu_L + \bar{e}_L\gamma_\mu e_L) = \frac{1}{2}Y_\mu. \quad (4.13)$$

The right hand side of (4.13) defines the *weak hypercharge* current, which has the property of commuting with the entire $SU(2)_L$ algebra . *Therefore, the minimal gauge group that contains the weak and the electromagnetic current is the product $SU(2)_L \otimes U(1)_Y$.*

It should be noted that only the electromagnetic current out of the four $SU(2)_L \otimes U(1)_Y$ currents is conserved. For the others the continuity equation is not satisfied because of the mass term of the electron, $\mathcal{L}_m$ in (4.10). For example, for the $SU(2)_L$ currents we find:

$$\partial^\mu(L^1_\mu + iL^2_\mu) = \partial^\mu(\bar{e}_L\gamma_\mu\nu_L) = im_e\bar{e}_R\nu_L \ ;$$

$$\partial^\mu L^3_\mu = i\frac{m_e}{2}(\bar{e}_L e_R - \bar{e}_R e_L) = -im_e\bar{e}\gamma_5 e \ , \quad (4.14)$$

where we have used the Dirac equation for ν and e.

The minimal $SU(2)_L \otimes U(1)_Y$ group contains two neutral generators. Consequently, we need to introduce a new neutral vector field, in addition to the photon, which must mediate interactions of a new type compared to the Fermi and electromagnetic interactions.

For completeness we summarise the classification of the fields according to the $SU(2)_L \otimes U(1)_Y$ quantum numbers:

$$l^e = \begin{pmatrix} (\nu_e)_L \\ e_L \end{pmatrix}_{-1} \ ; \quad (e_R)_{-2} \ . \quad (4.15)$$

For each multiplet we have indicated the value of the weak hypercharge. The field $(\nu_e)_R$ is completely neutral under $SU(2)_L \otimes U(1)_Y$ transformations. Therefore, in a Yang–Mills theory, it does not have any electroweak interaction and we can omit it completely from the electroweak Lagrangian.

From the scheme (4.15) we can obtain the covariant derivatives of the matter fields and therefore their interactions with the gauge fields of $SU(2)_L \otimes U(1)_Y$, which we denote as W^i_μ, $(i = 1, 2, 3)$ and B_μ respectively. We find:

$$D_\mu l^e = [\partial_\mu + ig\mathbf{W}_\mu \cdot \frac{\tau}{2} + ig'(-\frac{1}{2})B_\mu]l^e \ ,$$

$$D_\mu e_R = [\partial_\mu + ig'(-1)B_\mu]e_R \ , \quad (4.16)$$

and the interaction Lagrangian:

$$\mathcal{L}^{\nu,e}_{int} = -\bar{l}^e[g\mathbf{W}_\mu \cdot \frac{\tau}{2} + g'B_\mu(-\frac{1}{2})]\gamma^\mu l^e \ + \ g'\bar{e}_R\gamma^\mu e_R B_\mu \ . \quad (4.17)$$

The scheme (4.15) is repeated for the muon multiplet, (ν_μ, μ) and for the τ lepton, (ν_τ, τ).

Comment.   The problem of the existence of ν_R fields associated with e, μ and τ is still open. These fields could be absent completely, or they could be sensitive only to much weaker interactions, including gravity, and therefore be unobservable, at least for the moment. The possibility to insert the ν_R field in a wider scheme is discussed in chapter 9 of [1].

Problem.   Consider the charges associated with the chiral currents of (4.11), $L^i = \int d^3x L_0^i$, $R^i = \int d^3x R_0^i$, etc. Using the canonical anticommutation rules of the fields (cf. [1]), obtain the commutation rules of the algebra of $SU(2)_L \otimes SU(2)_R \otimes U(1)_{l_e} \otimes U(1)_A$, equation (4.7).

4.3   THE GLASHOW THEORY

As we saw in chapter 3, the Yang–Mills Lagrangian reduces, in the limit of zero coupling constant, to a Maxwell Lagrangian for each gauge field. In our case, cf. equation (3.38), we have:

$$\mathcal{L}_{YM}^{(g=0)} = -\frac{1}{4}W_{\mu\nu}^i W_{\mu\nu}^i - \frac{1}{4}B_{\mu\nu}^i B_{\mu\nu}^i \ . \tag{4.18}$$

To avoid the problem of the unwanted massless boson, we can, pragmatically, add a mass term which, following Glashow [10], we take to have the form:

$$\mathcal{L}_{gauge,mass} = \frac{1}{2}[M^2 \mathbf{W}_\mu \cdot \mathbf{W}^\mu + M_0^2 B_\mu B^\mu + 2M_{03}^2 W_\mu^3 B^\mu] \ . \tag{4.19}$$

In the case of charged fields ($i = 1, 2$), we define:

$$W_\mu = \frac{W_\mu^1 + iW_\mu^2}{\sqrt{2}}; \qquad W_\mu^\dagger = \frac{W_\mu^1 - iW_\mu^2}{\sqrt{2}} \ , \tag{4.20}$$

and we find:

$$
\begin{aligned}
\mathcal{L}_{gauge} = &-\frac{1}{2}W_{\mu\nu}^\dagger W^{\mu\nu} + M^2 W_\mu^\dagger \cdot W^\mu + \\
&-\frac{1}{4}[W_{\mu\nu}^3(W^3)^{\mu\nu} + B_{\mu\nu}B^{\mu\nu}] + \\
&+\frac{1}{2}[M^2 W_\mu^3(W^3)^\mu + M_0^2 B_\mu B^\mu + 2M_{03}W_\mu^3 B^\mu] \ .
\end{aligned}
\tag{4.21}
$$

Mass spectrum of the vector fields.   The first line of (4.21) defines two spin 1 bosons with electric charge ± 1 and mass M (cf. section. 5.6). For the neutral fields in the second and third lines, the physical fields (with definite mass) are identified by the eigenvectors of the *mass matrix* that, in the basis (W^3, B), is:

$$\mathcal{M} = \begin{pmatrix} M^2 & M_{03}^2 \\ M_{03}^2 & M_0^2 \end{pmatrix} \ . \tag{4.22}$$

This matrix is not completely arbitrary because it should have a zero eigenvalue, corresponding to the zero photon mass. We should therefore impose:

$$\det\mathcal{M} = 0; \quad \Rightarrow (M_{03}^2)^2 = M^2 M_0^2 \ .$$

We write the eigenvectors of the matrix (4.22) as:

$$Z_\mu = \cos\theta W_\mu^3 - \sin\theta B_\mu \ ,$$
$$A_\mu = \sin\theta W_\mu^3 + \cos\theta B_\mu \ , \tag{4.23}$$

where A_μ is the electromagnetic field and Z_μ is a new electrically neutral vector field. The non-zero eigenvalue of the mass matrix is simply given by its trace:

$$M_Z^2 = M^2 + M_0^2 \ . \tag{4.24}$$

On the other hand, the field A_μ in (4.23) should correspond to the zero eigenvalue, so we must have

$$\begin{pmatrix} \sin\theta, \cos\theta \end{pmatrix} \mathcal{M} \begin{pmatrix} \sin\theta \\ \cos\theta \end{pmatrix} = \sin^2\theta M^2 + 2\cos\theta\sin\theta M M_0 + \cos^2\theta M_0^2$$

$$= M^2 \left(\sin\theta + \cos\theta \frac{M_0}{M} \right)^2 = 0 \ ,$$

from which we find:

$$\frac{M_0^2}{M^2} = \tan^2\theta \tag{4.25}$$

and therefore:

$$M_Z^2 = \frac{M^2}{\cos^2\theta}. \tag{4.26}$$

Interactions with the physical fields. The inverse formulae to (4.23) are written:

$$W_\mu^3 = \cos\theta Z_\mu + \sin\theta A_\mu \ ;$$
$$B_\mu = -\sin\theta Z_\mu + \cos\theta A_\mu \ . \tag{4.27}$$

We can express the couplings of the interaction Lagrangian from equation (4.17) in the following way:

$$\mathcal{L}_{int} = -\bar{l}^e [g \frac{\tau^3}{2} W_\mu^3 + g'(-\frac{1}{2}) B_\mu] \gamma^\mu l^e + g' \bar{e}_R \gamma^\mu e_R B_\mu =$$

$$= -A_\mu [g \sin\theta(\frac{1}{2}\bar{\nu}_L \gamma^\mu \nu_L - \frac{1}{2}\bar{e}_L \gamma^\mu e_L) +$$

$$+ g' \cos\theta(\frac{1}{2}\bar{\nu}_L \gamma^\mu \nu_L - \frac{1}{2}\bar{e}_L \gamma^\mu e_L - \bar{e}_R \gamma^\mu e_R)] +$$

$$- Z_\mu [g \cos\theta(\frac{1}{2}\bar{\nu}_L \gamma^\mu \nu_L - \frac{1}{2}\bar{e}_L \gamma^\mu e_L)$$

$$- g' \sin\theta(-\frac{1}{2}\bar{\nu}_L \gamma^\mu \nu_L - \frac{1}{2}\bar{e}_L \gamma^\mu e_L - \bar{e}_R \gamma^\mu e_R)] \ . \tag{4.28}$$

To have the electromagnetic field correctly coupled to the electromagnetic current, we require:

$$g \sin \theta = g' \cos \theta = e \; ; \tag{4.29}$$

in particular:

$$\frac{g'}{g} = \tan \theta. \tag{4.30}$$

From (4.29) we find:

$$\mathcal{L}_{int} = -e A^\mu J_\mu^{e.m.} - \frac{g}{2 \cos \theta} Z^\mu J_\mu^Z - \frac{g}{2\sqrt{2}} [W_\mu (J^W)^\dagger \mu + \text{h.c.}] \tag{4.31}$$

where we have reinserted the charged boson coupling via the weak current in (4.1) and we have set:

$$J_\mu^{e.m.} = -(\bar{e}_L \gamma^\mu e_L + \bar{e}_R \gamma^\mu e_R);$$
$$J_\mu^Z = (\bar{\nu}_L \gamma^\mu \nu_L - \bar{e}_L \gamma^\mu e_L) + 2 \sin^2 \theta (\bar{e}_L \gamma^\mu e_L + \bar{e}_R \gamma^\mu e_R) =$$
$$= 2L_\mu^3 - 2 \sin^2 \theta J_\mu^{e.m.} \; . \tag{4.32}$$

The β decay amplitudes at low energy ($M_{fin} - M_{in} = Q << M$) are obtained to second order from (4.31) and they can be classified as follows:

- *Charged current* processes, due to the exchange of the charged boson, which give rise to the Fermi amplitudes:

$$A_{cc}(in \to fin) = \frac{G_F}{\sqrt{2}} \int d^4 x < fin | J_\mu^W(x) J_\mu^{W\dagger}(0) | in >$$

$$\frac{G_F}{\sqrt{2}} = \frac{g^2}{8M^2} = \frac{e^2}{8 \sin^2 \theta M^2} \; . \tag{4.33}$$

- *Neutral current* processes, due to the exchange of the Z boson, which give rise to amplitudes of the form:

$$A_{nc}(in \to fin) = \frac{g^2}{8M_Z^2} \int d^4 x < fin | J_\mu^Z(x) J_\mu^Z(0) | in > \; . \tag{4.34}$$

If we set

$$\frac{G_F^{nc}}{\sqrt{2}} = \frac{g^2}{8 \cos^2 \theta M_Z^2} \tag{4.35}$$

we see that, because of (4.26):

$$G_F^{nc} = G_F. \tag{4.36}$$

The result (4.36) is a notable consequence of the relation (4.26), which in its turn results from the form of (4.19). The scale of neutral current interactions is determined by the same Fermi constant which fixes the strength of the β interactions; the existence of neutral current processes with strength comparable to the Fermi interaction is a surprising but inevitable consequence of the form we have chosen to break the gauge symmetry with the masses of the intermediate bosons.

SPONTANEOUS SYMMETRY BREAKING

CONTENTS

In the first part of the chapter we consider a continuous global symmetry. The simplest case is that of a global symmetry under the $O(2)$ group considered by Goldstone. The extension to a gauge symmetry will be considered in the second part.

5.1   THE GOLDSTONE MODEL

We consider a complex scalar field with a quartic interaction. The Lagrangian has the form:

$$\mathcal{L} = \partial_\mu\phi\partial^\mu\phi^\dagger - \mu^2\phi\phi^\dagger - \lambda\left(\phi\phi^\dagger\right)^2 \tag{5.1}$$

from which we find the equation of motion:

$$(\Box + \mu^2)\phi + 2\lambda\phi(\phi\phi^\dagger) = 0. \tag{5.2}$$

The model has an exact symmetry under global phase changes of the field ϕ:

$$\phi'(x) = e^{i\alpha}\phi(x); \quad (\phi^\dagger)'(x) = e^{-i\alpha}\phi^\dagger(x) \tag{5.3}$$

where α is a real, arbitrary phase, the same at all points of space-time. The corresponding conserved current is calculated from Noether's theorem:

$$J^\mu = i\left[(\partial^\mu\phi)\phi^\dagger - \phi(\partial^\mu\phi^\dagger)\right]. \tag{5.4}$$

It is easy to verify, starting from the equation of motion (5.2), that the current is conserved:

$$\partial_\mu J^\mu = 0. \tag{5.5}$$

In many cases, it is useful to describe the theory using real fields, $\phi_{1,2}$, defined by:

$$\phi = \frac{\phi_1 + i\phi_2}{\sqrt{2}}; \quad \phi^\dagger = \frac{\phi_1 - i\phi_2}{\sqrt{2}} \ . \tag{5.6}$$

The Lagrangian and current take the form:

$$\mathcal{L} = \frac{1}{2}\left(\partial_\mu \phi_1 \partial^\mu \phi_1 + \partial_\mu \phi_2 \partial^\mu \phi_2\right) - V \ ;$$

$$V = \frac{\mu^2}{2}\left(\phi_1^2 + \phi_2^2\right) + \frac{\lambda}{4}\left(\phi_1^2 + \phi_2^2\right)^2 \ ;$$

$$J^\mu = \left[(\partial^\mu \phi_1)\phi_2 - \phi_1(\partial^\mu \phi_2)\right] \ . \tag{5.7}$$

For real fields, the transformation law (5.3) takes the infinitesimal form:

$$\delta\phi_i = -\alpha\epsilon_{ij}\phi_j \quad i,j = 1,2 \tag{5.8}$$

where ϵ_{ij} is the completely antisymmetric tensor in two dimensions with $\epsilon_{12} = -\epsilon_{21} = 1$ and other components zero. From (5.8) we recognise that the symmetry of the theory coincides with rotations around the origin of the Cartesian plane with axes ϕ_1 and ϕ_2, with a symmetry already evident in (5.7).

The energy density corresponding to the Lagrangian (5.1) is written:

$$\theta^{00} = \partial_0\phi\partial_0\phi^\dagger + (\nabla\phi) \cdot (\nabla\phi^\dagger) + V \ . \tag{5.9}$$

The derivative terms in θ^{00} are positive definite. The stability of the theory therefore requires that V should be a function *limited from below*. For the free theory, $\lambda = 0$, this requires that $\mu^2 > 0$. However, if $\lambda \neq 0$ the quartic term in the fields dominates at infinity and the condition that the Hamiltonian density has a lower limit is satisfied, provided that

$$\lambda > 0 \quad \text{(stability condition)}. \tag{5.10}$$

Thus, we have two possible theories according to the sign of the coefficient of $\phi\phi^\dagger$, which we will continue to denote as μ^2.

The classical field configuration which minimises the Hamiltonian (5.9) must be constant in space-time (for the derivative terms to vanish) and correspond to the absolute minimum of $V(\phi)$. The minimum energy configuration is invariant for space-time translations. Its quantum counterpart is the *vacuum state*, the state which contains no particles, the only invariant state for space-time translations. In the following, we will denote this state with:

$$|0\rangle : \quad \text{vacuum, or ground, state.} \tag{5.11}$$

We consider separately the two cases corresponding to the sign of μ^2.

1. $\mu^2 > 0$: The potential is a concave function of ϕ_1 and ϕ_2 with an absolute minimum at the origin:

$$V(\phi) = \text{minimum} = 0 \quad \text{for } \phi_1 = \phi_2 = 0. \tag{5.12}$$

In the limit $\lambda \to 0$ the Lagrangian (5.1) reduces to the Lagrangian of a complex Klein–Gordon field, ϕ, and is a combination of annihilation operators (of the particle of mass μ) and creation operators (of the antiparticle with the same mass). In each case, we have:

$$\langle 0|\phi(x)|0 \rangle = \langle 0|\phi(0)|0 \rangle = 0 \tag{5.13}$$

which is the quantum condition corresponding to (5.12).

The state of minimum energy of the field is unique and symmetric under the transformations (5.8). The particle spectrum is also symmetric; the symmetry (5.3) is realised exactly.

We do not know how to solve the theory for $\lambda > 0$. In the limit of small λ, perturbation theory produces a theory with scalar charged particles with interactions among themselves which are also symmetric for the transformations (5.3). If we can use this as guidance, we conclude that the case $\mu^2 > 0$ corresponds to the theory with *exact symmetry*:

$$\mu^2 > 0: \quad \langle 0|\phi(0)|0 \rangle = 0; \quad \text{exact symmetry.} \tag{5.14}$$

2. $\mu^2 < 0$: The form of the potential $V(\phi)$ in this case is illustrated in Figure 5.1.

The configuration with $\phi = 0$ is still an extremum of the potential but is not the configuration which minimises the potential. It corresponds, as shown in the figure, to a local maximum. The minimum of the potential is reached at all the points of a circle centred at the origin, which appears as the bottom of the valley in Figure 5.1. None of the points at the minimum is symmetric; the symmetry of the theory is reflected in the symmetry of the locations of the minima, which all correspond to the same value of the potential.

To resolve the degeneracy of the potential minima, we introduce a small additional term in the Lagrangian (5.1), a *driving term* characterised by a parameter ϵ, which is in general complex, that we will make tend to zero at the end. Therefore we consider the new Lagrangian:

$$\begin{aligned}
\mathcal{L}_\epsilon &= \partial_\mu \phi \partial^\mu \phi^\dagger - \mu^2 \phi \phi^\dagger - \lambda \left(\phi\phi^\dagger\right)^2 + \epsilon^*\phi + \epsilon\phi^\dagger = \\
&= \partial_\mu \phi \partial^\mu \phi^\dagger - \left[V(\phi) - \epsilon^*\phi - \epsilon\phi^\dagger\right]. \tag{5.15}
\end{aligned}$$

The energy minima are found by making the derivatives of the potential with respect to ϕ and $\phi^\dagger$ vanish. The value of ϕ at the minimum, which we denote with η_ϵ, is obtained from the equation:

$$\eta_\epsilon \left[\mu^2 + 2\lambda(\eta_\epsilon\eta_\epsilon^\dagger)\right] = \epsilon. \tag{5.16}$$

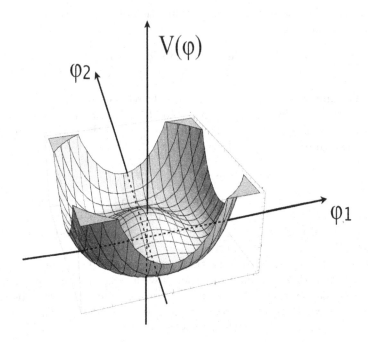

Figure 5.1 Higgs potential, $V(\phi)$, as a function of the real fields ϕ_1 and ϕ_2, for values $\mu^2 < 0$. The origin in field space, $\phi_1 = \phi_2 = 0$, corresponds to a maximum of the potential.

(The derivative with respect to ϕ provides the complex conjugate equation.)

For small values of ϵ, the equation has two roots, one close to the origin, $\eta_\epsilon = \mathcal{O}(\epsilon)$, which however corresponds to a maximum, the other with η_ϵ close to the circle of the minima of $V(\phi)$, which corresponds to the true minimum. We note that, in each case, the left hand side of (5.16) has the form $\eta_\epsilon \times$ (a real number), therefore η_ϵ must have the same phase as ϵ. The driving term forces the potential to have only one minimum in the ϕ_1–ϕ_2 plane, steering it in the direction which it itself determines. Taking ϵ real for simplicity, to first order in ϵ we have:

$$\eta_\epsilon = \eta + \delta;$$

$$\eta = \sqrt{\frac{-\mu^2}{2\lambda}};$$

$$\delta = -\frac{\epsilon}{2\mu^2}. \tag{5.17}$$

Note that the negative sign of μ^2 makes the square root real. In the limit

$\epsilon \to 0$, $\delta \to 0$ the minimum sits exactly on the circle of minima of $V(\phi)$, at the point $\phi_1 = \eta$, $\phi_2 = 0$.

In the quantum ground state, because of the usual correspondence between classical values and quantum expectation values, the field ϕ must have a non-zero expectation value, equal to:

$$\langle 0|\phi(0)|0\rangle = \eta. \tag{5.18}$$

The minimum energy configuration is not symmetric; the symmetry of the ground state is *spontaneously broken*:

$$\mu^2 < 0: \quad \langle 0|\phi(0)|0\rangle = \eta \neq 0; \quad \text{spontaneously broken symmetry.} \tag{5.19}$$

5.2   SPECTRUM OF FLUCTUATIONS FOR $\mu^2 < 0$

The small fluctuations around the ground state are described by oscillations of ϕ around its vacuum value. We set:

$$\phi(x) = \eta + \frac{\sigma_1(x) + i\sigma_2(x)}{\sqrt{2}}. \tag{5.20}$$

The functions $\sigma_i(x)$ are now the fields to quantise.

To determine the spectrum of particles associated with the small fluctuations, we must substitute the expression (5.20) in the Lagrangian given by (5.1) and expand in powers of σ_i. The masses are given by terms quadratic in the fields, while the higher order terms describe interactions between the particles. Hence, to find the masses, we can stop at second order terms of the expansion, which is given in a general form by the expression:

$$\mathcal{L}^{(2)} = \frac{1}{2} \sum_{i,j} \left[\frac{\partial^2 V(\bar{\phi}_1, \bar{\phi}_2)}{\partial \phi_i \partial \phi_j} \right] \sigma_i \sigma_j =$$

$$= \frac{1}{2} \sum_{i,j} M_{ij}^2 \sigma_i \sigma_j, \tag{5.21}$$

where the bar denotes the field values at the minimum point:

$$\bar{\phi}_1 = \sqrt{2}\eta, \quad \bar{\phi}_2 = 0. \tag{5.22}$$

Using the explicit form in (5.7), we find:

$$M_{11}^2 = \mu^2 + 6\lambda\eta^2 = -2\mu^2 = 4\lambda\eta^2$$
$$M_{12}^2 = M_{21}^2 = 0$$
$$M_{22}^2 = \mu^2 + 2\lambda\eta^2 = 0. \tag{5.23}$$

The two degrees of freedom originally associated with the complex field ϕ

(or with the real fields ϕ_1 and ϕ_2) are now represented by two particles with masses $4\lambda\eta^2$ and 0, respectively.

The appearance of a particle of exactly zero mass is the most surprising aspect of the result in (5.23). This result is known as *Goldstone's theorem* and the massless particle as the *Goldstone boson*.

We can easily convince ourselves that the presence of a massless particle is a general and inevitable consequence of spontaneous symmetry breaking. To do this, we note that the mass matrix in (12.2) is nothing other than the matrix of the curvature of the potential, calculated at the minimum point. Now, every time that the minimum is away from the origin, at a non-symmetric point, the symmetry requires that it should lie on a locus of equipotential points, in our case the circle $\phi_1^2 + \phi_2^2 = \eta^2 \neq 0$. But then there is a flat direction, the direction along the curve of the minimum, in which the second derivative is necessarily zero. The particle which corresponds to the oscillations of the field in this direction, σ_2 in our case, is therefore massless.

Comment 1. The phenomenon of spontaneous symmetry breaking is the basis of ferromagnetism. In this case, the symmetry in question is that of spatial rotations. The ground state of a ferromagnet breaks the symmetry since it determines a privileged direction in space, the direction in which the magnetism points spontaneously (the term 'spontaneous breaking' derives literally from this description). The oscillations orthogonal to the spontaneous magnetisation have the characteristic of having an energy which tends to zero in the long wavelength limit, or for frequencies which tend to zero; *energy* = *constant* × ν for $\nu \to 0$. This relation is the equivalent of the Goldstone theorem. Indeed, for a relativistic particle, *energy* = $\sqrt{m^2 + \nu^2}$ and the energy can be proportional to frequency only for zero mass.

If we cool a ferromagnet below the Curie temperature in the absence of an external magnetic field, we do not observe any spontaneous magnetisation. This is linked to the fact that the different magnetic domains, which still exist, are magnetised in randomly distributed directions and the spontaneous magnetisation averages to zero. The phenomenon is due to the fact that the different spatial directions are equivalent to each other. To have a spontaneous magnetisation it is necessary to cool the ferromagnet in the presence of a magnetic field, even weak, which orients all the domains in the same direction.

The Goldstone model has many analogies with the ferromagnet. In the model, the value of the field with minimum energy, $< 0|\phi|0 >$, takes the role of the spontaneous magnetisation of the ferromagnet. In the absence of a *driving term* the points on the circle of minimum energy in Figure 5.1 are all equivalent, like the directions of the magnetic domains. The *driving term* forces the vacuum, the ground state, to have a fixed direction in all of space-time.

Comment 2. Spontaneous breaking of a symmetry can happen only in systems with an infinite number of degrees of freedom, for example a ferromagnet of infinite volume or a field in an infinite volume, as in the Goldstone model. To illustrate this important fact, we consider, at the opposite extreme, a system with only one degree of freedom: a particle confined in perfectly symmetric double potential well [9]. In this case we have two candidates for the ground state: the states of the particle localised in the well to the right (ψ_1) or in that on the left (ψ_2). We have an example of degeneracy similar to the one encountered in Figure 5.1. The symmetry requires that:

$$< \psi_1|H|\psi_1 >= H_{11} =< \psi_2|H|\psi_2 >= H_{22} = H. \tag{5.24}$$

If we introduce a *driving term* which reduces the energy of one of the two wells, say number 1, the state ψ_1 becomes the ground state and the symmetry is broken.

However, to have a spontaneous breaking, the asymmetry must be maintained in the limit when the driving term tends to zero. This is not the case. Indeed, the Hamiltonian matrix between the two states also contains a non-diagonal term owing to the possibility that the particle could tunnel from one well to the other:

$$\begin{pmatrix} H & \delta \\ \delta & H \end{pmatrix}. \tag{5.25}$$

Since the two terms on the diagonal are equal, provided $\delta \neq 0$, no matter how small, the eigenstate of minimum energy of (5.25) will have the form:

$$\psi_0 = \frac{\psi_1 + \psi_2}{\sqrt{2}} \tag{5.26}$$

and the symmetry of the ground state is reestablished in the limit of zero driving term. The same thing holds, naturally, for systems with a finite number of degrees of freedom.

What happens for systems with infinite degrees of freedom is that the tunnelling amplitude from one state of minimum energy to another *can* be exactly zero (or better, tends to zero in the limit of infinite volume). In this case, once a driving term of strength ϵ takes the system into one of these energy minima, it remains there, as the quantum ground state, even in the limit $\epsilon \rightarrow 0$; the symmetry is spontaneously broken. We emphasise the word *can*. Indeed, systems with infinite degrees of freedom are known for both cases. For example, the Ising model in three or more dimensions has spontaneous breaking, but in two dimensions does not.

For a magnet of finite dimensions, the tunnelling amplitude of the magnetic orientation between one direction and another is not precisely zero. In these conditions, the magnet placed in a fixed direction should begin to precess until, after sufficient time, it ends up in an eigenstate symmetric for rotations. For a ferromagnet of macroscopic dimensions, however, the precession period can easily exceed the age of the universe. The approximation of considering the orientation of the magnet to be fixed is, in this case, more than adequate to describe the real situation.

5.3 GOLDSTONE'S THEOREM

The importance of Goldstone's theorem justifies a more general demonstration of the previous considerations, which we return to in this section.

The context is that of a Lagrangian theory of quantum fields, in which the Goldstone model of the previous section is a special case, for which the following conditions hold:

- The Lagrangian has an exact, continuous and global, symmetry (not extended to local symmetry through gauge fields); on the basis of Noether's theorem, a conserved current should therefore exist:

$$\partial_\mu J^\mu(x) = 0, \tag{5.27}$$

from which:

$$\frac{dQ}{dt} = 0; \qquad Q(t) = \int d^3x \; J^0(\mathbf{x}, t). \tag{5.28}$$

- Scalar fields, $\phi_i(x)$, exist, which are not invariant under the symmetry transformations; cf. (5.3).

- One of the fields ϕ_i, say ϕ_1, has a non-zero vacuum expectation value:

$$< 0|\phi_1(0)|0 >= \sqrt{2}\eta \neq 0. \tag{5.29}$$

Under these conditions the symmetry is spontaneously broken and we can prove what now follows.

The Goldstone theorem. A scalar particle, whose mass is exactly zero, exists, the Goldstone boson[1]. The particle is created from the vacuum by the scalar field and by the current. If we denote with $|p>$ the state of this particle with momentum p^μ, we have:

$$< p|\phi(0)|0 >= \frac{\sqrt{Z}}{\sqrt{(2\pi)^3 2\omega(p)}}; \tag{5.30}$$

$$< 0|J^\mu(0)|p >= \frac{ip^\mu \; F}{\sqrt{(2\pi)^3 2\omega(p)}}; \tag{5.31}$$

$$Z, \; F \neq 0. \tag{5.32}$$

---

[1]The Goldstone boson has spin zero and is a scalar particle under proper Lorentz transformations; from the point of view of parity, the Goldstone boson is a pseudoscalar if J^μ is an axial current.

The form of the matrix element of the field is the same as employed in Appendix A.2. For the form of the current matrix element, the argument is as follows. The matrix element $< 0|J^\mu(0)|p >$ must be proportional to a four-vector. On the other hand, the particle created by ϕ_2 from the vacuum has spin zero; therefore the state $|p >$ depends only on p^μ and the only 4-vector which can appear is p^μ. Hence the form indicated in (5.31) follows.

Preliminaries.   Without loss of generality, we can take for symmetry transformations those indicated in (5.8). In a quantum theory, symmetry transformations are represented by unitary operators of the form $exp(i\alpha Q)$ where the charge Q of equation (5.28) acts as an infinitesimal generator. Considering the transformation of ϕ_2 for infinitesimal α, we have:

$$(\phi_2)' = e^{-iQ\alpha}\phi_2 e^{iQ\alpha} = \phi_2 - i\alpha\,[Q, \phi_2] + \mathcal{O}(\alpha^2). \tag{5.33}$$

Comparing with the differential (5.8) we find the commutation relation:

$$i\frac{\delta\phi_2}{\delta\alpha} = [Q, \phi_2(x)] = \int d^3y\,[J^0(\mathbf{y}, t), \phi_2(\mathbf{x}, t)] = i\phi_1(\mathbf{x}, t). \tag{5.34}$$

For Q conserved, the time y^0 in which we evaluate J^0 is irrelevant and we have chosen $y^0 = x^0 = t$.

The central element in the proof of Goldstone's theorem is the Fourier transform of the current-field correlation function:

$$F^\mu(q) = \int d^4x\, e^{iqx} < 0|T\,[J^\mu(x)\phi_2(0)]\,|0 > . \tag{5.35}$$

We can repeat the considerations of Appendix A.2 concerning the two-point Green's function, and obtain for this correlation function a representation similar to the Kallen–Lehman representation. To this end, we separate in F^μ the contributions to the intermediate states of one particle, created from the vacuum by the application of ϕ_2, from those of two or more particles:

$$F^\mu(q) = F^\mu(q)_1 + F^\mu(q)_{>1} . \tag{5.36}$$

We first consider F_1^μ. For $x^0 > 0$, we have:

$$< 0|T\,[J^\mu(x)\phi_2(0)]\,|0 >_1 = \int d^3p\, e^{-ipx} < 0|J^\mu(0)|p >< p|\phi_2(0)|0 >=$$

$$= F\sqrt{Z} \int \frac{d^3p}{(2\pi)^3 2\omega(p)}\, e^{-ipx}\, ip^\mu \tag{5.37}$$

where we have used the definitions (5.30) and (5.31) for the matrix elements.

Reasoning as we did in the case of the Green's function, it is easily seen that, with the definition of the additional factors given in (5.32), the variable F is Lorentz invariant, therefore depends only on p^2, i.e. on the mass of the particle in $|p >$ and is therefore a constant.

From (5.37), adding the case $x^0 < 0$, we immediately obtain:

$$F^\mu(q)_1 = F \sqrt{Z} \, iq^\mu \tilde{D}_F(q, m) \qquad (5.38)$$

where $\tilde{D}$ is the Fourier transform of the propagator of the intermediate particle, to which we attribute a mass m, for the moment not determined.

We leave to the reader the task of proving that the contribution of the states of two or more particles is of the form:

$$F^\mu(q)_{>1} = q^\mu \int_{M_0^2}^{+\infty} \rho(M^2) \, \frac{dM^2}{q^2 - M^2} \qquad (5.39)$$

where M_0 is the minimum value of the invariant mass of these states.

Proof. The sought-after result is obtained by demanding that $F^\mu(q)$ satisfies the conditions required by conservation of the current, equation (5.27), and from the non-zero value of the field in the ground state, (5.29).

Multiplying both sides of (5.35) by q_μ, we obtain:

$$q_\mu F^\mu(q) = q_\mu \int d^4x \, e^{iqx} < 0| T\left[J^\mu(x)\phi_2(0)\right] |0> =$$

$$= \int d^4x \left(-i\partial_\mu e^{iqx}\right) < 0| T\left[J^\mu(x)\phi_2(0)\right] |0> =$$

$$= i \int d^4x \, e^{iqx} \, \partial_\mu < 0| T\left[J^\mu(x)\phi_2(0)\right] |0> =$$

$$= i \int d^4x \, e^{iqx} < 0| T\left[\partial_\mu J^\mu(x)\phi_2(0)\right] |0> +$$

$$+ i \int d^4x \, e^{iqx} \delta(x^0) < 0| \left[J^0(\mathbf{x}, 0)\phi_2(0)\right] |0> =$$

$$= i \int d^3x \, e^{-i\mathbf{q}\mathbf{x}} < 0| \left[J^0(\mathbf{x}, 0)\phi_2(0)\right] |0> . \qquad (5.40)$$

If now we take the limit $q_\mu \to 0$ and use (5.34) and (5.29), we obtain the relation[2]:

$$\lim_{q\to 0} q_\mu F^\mu(q) = i < 0| \left[Q, \phi_2(0)\right] |0> = - < 0|\phi_1(0)|0> = -\sqrt{2}\eta \neq 0. \quad (5.41)$$

If we neglect the contribution to the left hand side of the intermediate states with two or more particles (we will see in a moment that this term vanishes in the limit), we obtain:

$$\lim_{q\to 0} q_\mu F^\mu(q)_1 = \sqrt{Z} F \lim_{q\to 0} q^2 \tilde{D}_F(q, m) = -\sqrt{2}\eta. \qquad (5.42)$$

---

[2]In the literature, equations of this type, which are derived from the conservation of the current, are known as *Ward identities*.

This requires that, in the neighbourhood of $q^2 = 0$, we should have:

$$\tilde{D}_F(q, m) = \frac{-\sqrt{2}\eta}{\sqrt{Z}F}\frac{1}{q^2} \quad (q^2 \text{ near to } 0). \tag{5.43}$$

This equation, to be satisfied, requires:

- $m = 0$, value for which:

$$\tilde{D}_F(q, m) = \frac{1}{q^2}, \tag{5.44}$$

- and

$$\sqrt{Z}F = -\sqrt{2}\eta. \tag{5.45}$$

In other words, ϕ_2 must create from the vacuum a massless particle, the Goldstone boson, and both Z and F must be non-zero.

To complete the proof, we must consider the contribution of $F^\mu(q)_{>1}$ to (5.41).

From (5.39), we obtain:

$$q_\mu F^\mu(q)_{>1} = q^2 \int_{M_0^2}^{+\infty} \rho(M^2)\frac{dM^2}{q^2 - M^2}. \tag{5.46}$$

In the limit $q^2 \to 0$ only the singular part of the integral counts on the right hand side, and can exist if $M_0 = 0$. To separate the singularity, we set:

$$\int_0^{+\infty} \rho(M^2)\frac{dM^2}{q^2 - M^2} = \int_0^{\bar{M}^2} \rho(M^2)\frac{dM^2}{q^2 - M^2} + \int_{\bar{M}^2}^{+\infty} \rho(M^2)\frac{dM^2}{q^2 - M^2} \tag{5.47}$$

and we choose $\bar{M} > 0$, but small enough that the variation of $\rho(M^2)$ in the interval $(0, \bar{M}^2)$ can be ignored. The first term can be calculated explicitly[3]:

$$\int_0^{\bar{M}^2} \rho(M^2)\frac{dM^2}{q^2 - M^2} = \rho(0)\int_0^{\bar{M}^2}\frac{dM^2}{q^2 - M^2} = \rho(0)\log(\frac{q^2}{q^2 - \bar{M}^2}). \tag{5.48}$$

As can be seen, the singularity in $q^2 = 0$ is only logarithmic, therefore:

$$\lim_{q^2 \to 0} q_\mu F^\mu(q)_{>1} = 0 \tag{5.49}$$

as expected.

---

[3]To remain in the region of analyticity of $F_{>1}$ we must let $q^2 \to 0$ from negative values, for which the logarithm which appears in (5.48) is well defined.

Note 1. As a result of the conditions (5.32), the Goldstone boson is fully observable in the reactions predicted by the theory. On the other hand, massless scalar relativistic particles have never been observed in subnuclear physics. Goldstone's theorem was for a long time the major obstacle to interpret the observed symmetry violations, for example of isotopic spin, as due to spontaneous symmetry breaking.

Note 2. We have not used the fact that ϕ_i are canonical elementary fields, as happens in the Goldstone model. The theorem also holds in the case in which the scalar field that has a vacuum expectation value is a general local operator, for example the product of fundamental fields. This situation is encountered in quantum chromodynamics, the theory of quarks and gluons, in connection with the spontaneous breaking of the chiral symmetry, see chapter 11.

5.4 THE BROUT–ENGLERT–HIGGS MECHANISM

We return to the model of the beginning of this chapter, and promote the symmetry of global phase transformations to a gauge symmetry [11, 12]. This requires the introduction of a vector field A^μ similar to the electromagnetic field, as discussed in chapter 3. The resulting Lagrangian is:

$$\mathcal{L} = (D_\mu \phi)^\dagger D^\mu \phi - V(\phi) - \frac{1}{4} F_{\mu\nu} F^{\mu\nu};$$
$$V(\phi) = \mu^2 \phi^\dagger \phi + \lambda (\phi^\dagger \phi)^2 - \epsilon \phi^\dagger - \epsilon^* \phi;$$
$$D^\mu = \partial^\mu - ieA^\mu;$$
$$F^{\mu\nu} = \partial^\nu A^\mu - \partial^\mu A^\nu. \qquad (5.50)$$

In the limit $\epsilon \to 0$, $\mathcal{L}$ is invariant under the gauge transformations:

$$\phi(x) \to e^{i\alpha(x)} \phi(x); \qquad \phi(x)^\dagger \to e^{-i\alpha(x)} \phi(x)^\dagger; \qquad (5.51)$$
$$A^\mu \to A^\mu + \frac{1}{e} \partial^\mu \alpha(x). \qquad (5.52)$$

where $\alpha(x)$ is an arbitrary real function of x and e is a new coupling constant, which coincides with the electric charge if we identify A^μ with the electromagnetic field. As before we should have $\lambda > 0$ to obtain a stable theory, but μ^2 can be either negative or positive.

1. $\mu^2 > 0$. The minimum energy configuration corresponds to $\phi = 0$, $A^\mu = 0$. If we quantize ϕ and A^μ we obtain a theory with:

- a charged particle and its antiparticle, both with mass $\mu \neq 0$,

- a massless spin 1 particle with two polarisation states, very similar to the photon.

For sufficiently small λ and e, the Lagrangian describes interactions of scalar particles with the electromagnetic field (scalar electrodynamics, coupling constant e) and their self-interactions (coupling constant λ).

2. $\mu^2 < 0$.   The minimum energy configuration occurs for $A^\mu = 0$, to minimise the electrostatic energy; therefore we revert to the case of the Goldstone model, for which (with real $\epsilon \to 0$):

$$V(\bar{\phi}) = \text{minimum}: \quad \bar{\phi} = \eta = \sqrt{\frac{-\mu^2}{2\lambda}} + \mathcal{O}(\epsilon). \qquad (5.53)$$

To study the fluctuations around the minimum configuration, we again set:

$$\phi = \eta + \frac{\sigma_1(x) + i\sigma_2(x)}{\sqrt{2}} \qquad (5.54)$$

and expand the Lagrangian in (5.50). The mass spectrum of associated particles is obtained as before from the terms quadratic in the fields σ_i and A^μ. However, before considering these terms, we note that under the transformations (5.52) with infinitesimal α the field ϕ transforms as:

$$\phi \to \phi + i\alpha\phi = \eta + \frac{\sigma_1(x) + i\sigma_2(x)}{\sqrt{2}} + i\eta\alpha(x) \qquad (5.55)$$

or:

$$\sigma_1(x) \to \sigma_1'(x) = \sigma_1(x); \quad \sigma_2(x) \to \sigma_2'(x) = \sigma_2(x) + \sqrt{2}\eta\alpha(x) \qquad (5.56)$$

where we have treated both σ_i and α as first order quantities and neglected terms of second and higher order. Naturally, at the same time we must transform A^μ according to (5.52).

The field σ_2 transforms in a non-homogeneous way because of the addition of a term proportional to α but, differently from the global case, the additional term is a *function of x* which we can arrange as we please. In particular, given the fields $\sigma_i(x)$ and $A^\mu(x)$, we can choose:

$$\alpha(x) = -\frac{\sigma_2(x)}{\sqrt{2}\eta} \qquad (5.57)$$

so as to have:

$$\sigma_2'(x) = 0 \quad \text{(unitary gauge)}. \qquad (5.58)$$

The field σ_2, which in the model with global symmetry corresponds to the Goldstone boson, can be completely eliminated in the case of gauge symmetry and therefore does not correspond to any physical degree of freedom. The gauge identified by condition (5.58) is commonly known as the *unitary gauge*.

5.5   MASSES OF VECTOR PARTICLES IN THE UNITARY GAUGE

In the unitary gauge, in which

$$\phi = \eta + \frac{\sigma(x)}{\sqrt{2}} = \rho(x) = \text{real} \tag{5.59}$$

the Lagrangian takes the form:

$$\mathcal{L} = \frac{1}{2}\partial_\mu \sigma \partial^\mu \sigma + e^2 \rho(x)^2 A_\mu A^\mu - V(\rho) - \frac{1}{4}F_{\mu\nu}F^{\mu\nu} \tag{5.60}$$

and it is straightforward to extract the terms quadratic in the fields. Recalling the first of (5.23), we find

$$\mathcal{L} = \frac{1}{2}\partial_\mu \sigma \partial^\mu \sigma - \frac{1}{2}M_H^2 \sigma^2 - \frac{1}{4}F_{\mu\nu}F^{\mu\nu} + \frac{1}{2}M_A^2 A_\mu A^\mu \tag{5.61}$$

with:

$$M_H^2 = -2\mu^2 = 4\lambda\eta^2;$$
$$M_A^2 = 2e^2\eta^2. \tag{5.62}$$

The Lagrangian describes a neutral scalar particle of mass M_H and, as we will show explicitly in Section 5.6, a spin 1 particle with mass M_A. We can summarise the results found in the following way:

- In the presence of a local (gauge) symmetry the spontaneous breaking of the symmetry implies that the corresponding gauge field acquires a non-zero mass.

The field corresponding to the Goldstone boson in the global theory (which can be eliminated by a gauge transformation) disappears. In its place the degree of freedom missing from the vector field appears to create the field of a spin 1 particle with mass. It is important to note that in the transition from an exact symmetry to a broken symmetry ($\mu^2 < 0$) the mass spectrum changes dramatically but the number of degrees of freedom is conserved, as shown in Table 5.1.

Table 5.1   Mass spectrum of the particles in the Higgs model.

–	σ_1	σ_2	A^μ	no. degrees of freedom
$\mu^2 > 0$	μ	μ	0	1+1+2=4
$\mu^2 < 0$	M_H	–	M_A	1+0+3=4

From a physical point of view, the *Brout–Englert–Higgs mechanism* eliminates the difficulty connected to the conspicuous absence of massless particles in subnuclear physics, both scalar, according to the Goldstone theorem, or vector, in relation to a Yang–Mills theory. This result opens the path to a theory of the weak force mediated by massive intermediate vector bosons.

The only known massless particle is the photon, corresponding to the gauge symmetry of the electromagnetic interactions which, within extraordinarily precise limits, is the only gauge symmetry respected by Nature.

5.6 MASSIVE VECTOR FIELD

In completion of the arguments of the previous section, we show that the Lagrangian of the vector field in equation (5.61) corresponds to a massive spin 1 particle.

We start from the Lagrangian:

$$\mathcal{L}_A = -\frac{1}{4}F_{\mu\nu}F^{\mu\nu} + \frac{1}{2}M^2 A_\mu A^\mu. \tag{5.63}$$

The most direct way of obtaining the result is to calculate the classical Green's function which, with the Feynman prescription for the integration path, provides the quantum propagator.

We first determine the equation of motion. From (5.63) we easily find:

$$\frac{\partial \mathcal{L}}{\partial \partial_\alpha A_\beta} = F_{\alpha\beta};$$

$$\frac{\partial \mathcal{L}}{\partial A_\beta} = M^2 A_\beta; \tag{5.64}$$

from which the Euler–Lagrange equations are obtained:

$$\partial^\alpha F_{\alpha\beta} = M^2 A_\beta, \tag{5.65}$$

or

$$-\Box A_\beta + \partial_\beta \left(\partial^\alpha A_\alpha\right) - M^2 A_\beta = 0. \tag{5.66}$$

The corresponding Green's function, $G_{\mu\nu}$, is the solution of the equation:

$$\left[\left(-\Box - M^2\right)\delta^\rho_\mu + \partial^\rho \partial_\mu\right] G_{\rho\nu}(x) = -g_{\mu\nu}\delta^{(4)}(x). \tag{5.67}$$

Taking the Fourier transform, we obtain the equation for $\tilde{G}$:

$$K^\rho_\mu(k)\tilde{G}_{\rho\nu}(k) = \left[\left(k^2 - M^2\right)\delta^\rho_\mu - k^\rho k_\mu\right]\tilde{G}_{\rho\nu}(k) = -g_{\mu\nu}. \tag{5.68}$$

Compared to the analogous equation for the electromagnetic field, equation (5.68) has in addition the mass term, which makes the matrix $K(k)$ invertible[4]. The solution of (5.68) is found just by inverting K, which must have the form:

$$(K^{-1})^\nu_\rho(k) = A(k^2)\delta^\nu_\rho + B(k^2)k^\nu k_\rho$$

---

[4]In the electromagnetic case, the matrix K is substituted by $H^\rho_\mu(k) = k^2\delta^\rho_\mu - k^\rho k_\mu$. H is not invertible, since it has an eigenvector with zero eigenvalue: $k^\mu H^\rho_\mu(k) = 0$.

with A and B Lorentz invariant functions of k^μ, hence functions of k^2. Imposing:

$$K_\mu^\rho(k)(K^{-1})_\rho^\nu(k) = \delta_\mu^\nu \qquad (5.69)$$

we find:

$$A(k^2) = \frac{1}{k^2 - M^2}$$
$$B(k^2 - M^2) - Bk^2 - A = 0, \qquad (5.70)$$

from which, finally:

$$\tilde{G}^{\mu\nu}(k) = -(K^{-1})^{\mu\nu} = \frac{-g^{\mu\nu} + \frac{k^\mu k^\nu}{M^2}}{k^2 - M^2 + i\epsilon} \ . \qquad (5.71)$$

The pole at $k^2 = M^2$ shows that the field corresponds to a particle of mass M. If we set ourselves at the pole and in the rest system of the particle, $k^0 = M$ and $\vec{k} = 0$, the non-vanishing components of the numerator in (5.71) are those with spatial indices. We obtain:

$$-g^{\mu\nu} + \frac{k^\mu k^\nu}{M^2} \to \delta_{ij} = P_{ij} \ . \qquad (5.72)$$

On the pole, the numerator of the propagator is, in general, the projection operator on the states of the particle and its trace directly gives the number of available states, equal to $2S + 1$, where S is the spin. Since $\text{Tr}P = 3$, it follows that $S = 1$.

CHAPTER 6

ELECTROWEAK UNIFICATION II

CONTENTS

6.1 THE THEORY OF WEINBERG AND SALAM

The starting point is the theory based on the symmetry $SU(2)_L \otimes U(1)_Y$ illustrated earlier in section 4.3, in its exactly symmetric version, i.e. without *ad hoc* mass terms for the vector and electron fields. The Lagrangian follows from the classification under $SU(2)_L \otimes U(1)_Y$ of the lepton fields given in (4.15) that we repeat for convenience:

$$l = \begin{pmatrix} (\nu_e)_L \\ e_L \end{pmatrix}_{Y=-1} ; \quad (e_R)_{Y=-2} . \tag{6.1}$$

The corresponding Yang–Mills Lagrangian is therefore:

$$\mathcal{L}_{eW} = \bar{l}i\gamma^\mu D_\mu l + \bar{e}_R i\gamma^\mu D_\mu e_R - \frac{1}{4}\left[\mathbf{W}_{\mu\nu}\mathbf{W}^{\mu\nu} + B_{\mu\nu}B^{\mu\nu}\right]. \tag{6.2}$$

Covariant derivatives and field tensors are given by:

$$D_\mu l = [\partial_\mu + ig\mathbf{W}_\mu \cdot \frac{\tau}{2} + ig'(-\frac{1}{2})B_\mu]l ;$$
$$D_\mu e_R = [\partial_\mu + ig'(-1)B_\mu]e_R ;$$
$$W^i_{\mu\nu} = \partial_\nu W^i_\mu - \partial_\mu W^i_\nu + g\,\epsilon_{jik}W^j_\mu W^k_\nu ;$$
$$B_{\mu\nu} = \partial_\nu B_\mu - \partial_\mu B_\nu . \tag{6.3}$$

At this stage, the theory describes fermions and vector fields, *all massless.*

6.2   THE SCALAR DOUBLET

We now introduce a scalar field which can trigger the symmetry breaking, conserving the gauge symmetry of electromagnetism, according to the scheme:

$$SU(2)_L \otimes U(1)_Y \to U(1)_{em} . \tag{6.4}$$

We have little information and several possibilities for the scalar field. The choice of Weinberg and Salam allows the spontaneous breaking mechanism also to generate the masses of the electron and the quarks, in the extension of the theory to hadronic particles, cf. chapter 10, to give us a completely realistic theory. The choice in question consists of introducing an $SU(2)_L$ doublet with $Y = +1$:

$$\phi = \begin{pmatrix} \phi^+ \\ \phi^0 \end{pmatrix}_{Y=+1} \quad ;$$

$$D_\mu \phi = [\partial_\mu + ig\mathbf{W}_\mu \cdot \frac{\tau}{2} + ig'(+\frac{1}{2})B_\mu]\phi . \tag{6.5}$$

We must add the part of the Higgs doublet, also perfectly symmetric, to the Lagrangian in (6.2):

$$\mathcal{L}_{tot} = \mathcal{L}_{eW} + \mathcal{L}_{\phi W} \tag{6.6}$$

with

$$\mathcal{L}_{\phi W} = (D_\mu \phi)^\dagger (D^\mu \phi) - V(\phi);$$
$$V(\phi) = \mu^2 \phi^\dagger \phi + \lambda(\phi^\dagger \phi)^2 . \tag{6.7}$$

We can now suppose, as in section 5.4, that ϕ has a vacuum expectation value, thus breaking the symmetry. Within redefinitions, we can always suppose that the lower component should be the non-vanishing one:

$$\bar\phi = < 0|\phi|0 >= \begin{pmatrix} 0 \\ \eta \end{pmatrix} \quad ;$$

$$\eta = \sqrt{\frac{-\mu^2}{2\lambda}} . \tag{6.8}$$

For the doublet (6.5) the electric charge is represented by the matrix:

$$Q = \begin{pmatrix} +1 & 0 \\ 0 & 0 \end{pmatrix} . \tag{6.9}$$

The symmetry breaking induced by $\bar\phi \neq 0$ realises the scheme (6.4), since the field at the minimum is invariant under the phase transformations associated with the group $U(1)_{em}$:

$$e^{i\alpha Q}\bar\phi = \begin{pmatrix} e^{i\alpha} & 0 \\ 0 & 1 \end{pmatrix} \begin{pmatrix} 0 \\ \eta \end{pmatrix} = \bar\phi . \tag{6.10}$$

To correctly identify the physical particles predicted by the theory we must identify the conditions for the unitary gauge. To do this, we note that every two-dimensional spinor can be reduced to a spinor with just a *lower* and *real* component by a transformation of the symmetry group, dependent on location. In formulae, any $\phi(x)$ of the form (6.5) can be put in the standard form:

$$\phi(x) = U(x) \begin{pmatrix} 0 \\ \rho(x) \end{pmatrix} \tag{6.11}$$

with $\rho(x)$ real and $U(x)$ a matrix of $SU(2)_L \otimes U(1)_Y$.

Proof.   We consider a general spinor:

$$\phi = \begin{pmatrix} z_1 \\ z_2 \end{pmatrix}$$

with complex $z_{1,2}$. Multiplying the spinor by the diagonal matrix:

$$e^{i(\alpha I_3 + \beta Y)} = \begin{pmatrix} e^{i\alpha} e^{i\beta} & 0 \\ 0 & e^{-i\alpha} e^{i\beta} \end{pmatrix} \tag{6.12}$$

we can eliminate the phases of z_1 and z_2 and make the spinor purely real. At this point we apply an $SU(2)_L$ matrix corresponding to a rotation about axis-2:

$$U_2(\gamma) = e^{i\gamma \frac{\tau_2}{2}} \ . \tag{6.13}$$

With an appropriate choice of γ, we can eliminate the *upper* component of the spinor, keeping the *lower* component real, and we obtain the standard form:

$$U = U_2(\gamma) e^{i(\alpha I_3 + \beta Y)} \quad (U \text{ belongs to } SU(2)_L \otimes U(1)_Y). \tag{6.14}$$

We subject all the fields to the gauge transformation corresponding to $U(x)^{-1}$. The Lagrangian remains invariant and the Higgs field takes the real form which characterises, in the case in question, the unitary gauge:

$$\phi(x) = \begin{pmatrix} 0 \\ \rho(x) \end{pmatrix} \quad (\rho(x) \text{ real}). \tag{6.15}$$

In general, in this gauge, we can write:

$$\phi(x) = \begin{pmatrix} 0 \\ \eta + \frac{\sigma(x)}{\sqrt{2}} \end{pmatrix} \quad (\text{unitary gauge}). \tag{6.16}$$

In the Higgs doublet, only a single physical field, $\sigma(x)$, remains and hence a neutral scalar particle, the *Higgs boson*.

6.3 MASSES OF THE VECTOR FIELDS

As in the Higgs model, the masses of the vector fields have their origin in the covariant derivative terms of $\mathcal{L}_{\phi W}$. In the gauge (6.15), the Lagrangian becomes:

$$\mathcal{L}_{\phi W} = \frac{1}{2}\partial_\mu \sigma \partial^\mu \sigma - V\left[\eta + \frac{\sigma(x)}{\sqrt{2}}\right] +$$
$$+ g^2 W_\mu^i (W^j)^\mu \left[\bar{\phi}\frac{\tau_i \tau_j}{4}\bar{\phi}\right] + (g')^2 \frac{1}{4}\eta^2 B_\mu B^\mu + 2gg' W_\mu^3 B^\mu \left[\bar{\phi}\frac{\tau_3}{4}\bar{\phi}\right] \quad (6.17)$$

(with summation over repeated indices). Using equation (6.8) and the properties of the Pauli matrices, we find:

$$W_\mu^i (W^j)^\mu \left[\bar{\phi}\frac{\tau_i \tau_j}{4}\bar{\phi}\right] = \frac{1}{4}\eta^2 \mathbf{W}_\mu \mathbf{W}^\mu;$$
$$W_\mu^3 B^\mu \left[\bar{\phi}\frac{\tau_3}{4}\bar{\phi}\right] = -\frac{1}{4}\eta^2 W_\mu^3 B^\mu. \quad (6.18)$$

Comparing with section 4.3 we see that we have again obtained equation (4.19) with:

$$M^2 = \frac{1}{2}g^2 \eta^2;$$
$$M_0^2 = \frac{1}{2}(g')^2 \eta^2;$$
$$M_{03}^2 = -\frac{1}{2}gg'\eta^2; \quad (6.19)$$

or:

$$\mathcal{M} = \frac{1}{2}\eta^2 \begin{pmatrix} g^2 & -gg' \\ -gg' & (g')^2 \end{pmatrix}. \quad (6.20)$$

Comparing with chapter 4 we see that M is the mass of the charged intermediate boson. From (6.19) and (4.33) we can connect η directly to the Fermi constant:

$$\eta^{-2} = \frac{g^2}{2M^2} = \frac{4G_F}{\sqrt{2}} = 2\sqrt{2}G_F; \quad (6.21)$$
$$\eta \simeq 174 \text{ GeV}. \quad (6.22)$$

As we have seen, the vacuum configuration is invariant under gauge transformations associated with electric charge. Therefore it is not surprising to notice that the matrix (6.20) automatically satisfies the condition of having zero determinant, $det\mathcal{M} = 0$, hence allowing a massless photon.

As in section 4.3, we denote by Z^μ the field with mass and A^μ the electromagnetic field which together diagonalise (6.20), and we use the same convention for the electroweak mixing angle, θ.

$$Z_\mu = \cos\theta W_\mu^3 - \sin\theta B_\mu;$$
$$A_\mu = \sin\theta W_\mu^3 + \cos\theta B_\mu. \quad (6.23)$$

The condition that the field A^μ should be the eigenvector of (6.20) with zero eigenvalue is written:

$$0 = \begin{pmatrix} g^2 & -gg' \\ -gg' & (g')^2 \end{pmatrix} \begin{pmatrix} \sin\theta \\ \cos\theta \end{pmatrix} = \begin{pmatrix} g^2\sin\theta - gg'\cos\theta \\ -gg'\sin\theta + (g')^2\cos\theta \end{pmatrix}. \quad (6.24)$$

The vanishing of the right hand side requires:

$$\tan\theta = \frac{g'}{g} \quad (6.25)$$

which is precisely the condition for which the field A^μ defined in (6.23) couples to the electron via the electromagnetic current with:

$$g\sin\theta = g'\cos\theta = e; \quad (6.26)$$

as can be seen from equations (4.30) and (4.29).

In conclusion, the spontaneous breaking in the theory of Weinberg and Salam, based on an exactly symmetric Lagrangian, completely reproduces the mass spectrum and the coupling of the vector fields obtained from the theory of Glashow, in particular the results summarised in (4.26) and (4.36), and the form of the neutral current coupled to Z^μ, (4.32), results which, as we will see, are in excellent agreement with experimental data.

6.4 MASS OF THE ELECTRON

To complete the electroweak theory we must account for the mass of the electron, which in the Glashow theory is described by the term $\mathcal{L}_m$ which explicitly breaks the symmetry, cf. equation (4.10):

$$\mathcal{L}_m = m_e \bar{e}e = m_e(\bar{e}_L e_R + \bar{e}_R e_L). \quad (6.27)$$

The reason for the non-invariance of $\mathcal{L}_m$ under transformations of $SU(2)_L \otimes U(1)_Y$ is explained by the form of the right hand side of (6.27): the field $\bar{e}_L$ has weak isospin $\frac{1}{2}$ while e_R has isospin zero (see equation (6.1)), and hence $\mathcal{L}_m$ in the sum has weak isospin $\frac{1}{2}$. Following the spontaneous breaking, ϕ acquires a constant component, which reproduces the Lagrangian $\mathcal{L}_m$ while the quantum component of ϕ gives rise to a new interaction between ϕ and the electron.

As a formula, we write the invariant Lagrangian:

$$\mathcal{L}_{e\phi} = g_e \left(\bar{l}\phi e_R + \bar{e}_R \phi^\dagger l \right). \quad (6.28)$$

The explicit expansion of the invariant term $\bar{l}\phi$ in the presence of spontaneous symmetry breaking, in the unitary gauge (6.16), is:

$$\bar{l}\phi = \bar{\nu}_L \phi^+ + \bar{e}_L \phi^0 = \bar{e}_L(\eta + \frac{\sigma}{\sqrt{2}}) \quad (6.29)$$

from which we obtain:

$$\mathcal{L}_{e\phi} = g_e \eta \bar{e}e + g_e \frac{\sigma}{\sqrt{2}} \bar{e}e = \mathcal{L}_m + \text{interaction}. \tag{6.30}$$

We have obtained a mass term for the electron with:

$$m_e = g_e \eta. \tag{6.31}$$

It can be observed that, after the spontaneous breaking, the symmetry has not been lost. A relation remains between the mass of the electron and the constant of the interaction between the electron and the Higgs boson which we can rewrite:

$$g_e = \frac{m_e}{\eta} = \left(2\sqrt{2}G_F\right)^{1/2} m_e. \tag{6.32}$$

The interaction is determined by the Fermi constant and the electron mass. The relation (6.32) completely characterises the interaction of the Higgs boson with the electron and provides a characteristic signature for its identification.

6.5 MASS OF THE NEUTRINO

The combination which appears in (6.29) is not the only invariant we can construct with the *left handed* electron doublet and the Higgs doublet. Using the invariant antisymmetric tensor in two dimensions[1] the invariant combination is obtained:

$$\left(l^i \phi^j\right) \epsilon_{ij} = \nu_L \phi^0 - e_L \phi^+ \quad (Y = -2). \tag{6.33}$$

In the presence of a field ν_R we can obtain a Lagrangian which, after spontaneous breaking, generates a mass for the neutrino:

$$\mathcal{L}_{\nu\phi} = g_D \left[\bar{\nu}_R \left(l^i \phi^j\right) \epsilon_{ij} + \text{h.c.}\right] =$$
$$= g_D \eta \left(\bar{\nu}_R \nu_L + \bar{\nu}_L \nu_R\right). \tag{6.34}$$

The possible mass terms for the neutrino were characterised in chapter 13 of [1]. According to the terminology adopted there, the term in (6.34) is a *Dirac mass*.

Comment: the seesaw mechanism. The problem of (6.34) is that it is difficult to attribute to the same mechanism, i.e. the spontaneous breaking of $SU(2)_L \otimes U(1)_Y$, the generation of masses as different as those of the electron ($m_e \simeq 0.5$ MeV) and ν_e (where it is possible that $m_{\nu_e} \simeq 10^{-4}$ eV, ten orders of magnitude smaller).

---

[1] $\epsilon_{ij} = -\epsilon_{ji}$, $\epsilon_{12} = +1$.

However, with the field ν_R we can construct a *Majorana mass* which is invariant under $SU(2)_L \otimes U(1)_Y$, of the form:

$$\mathcal{L}_{nu-M} = M \left(\nu_R \gamma^0 \nu_R + \text{h.c.}\right). \tag{6.35}$$

The neutrino mass arises from the combination of these two terms:

$$\begin{aligned}\mathcal{L}_{\nu-tot} = \mathcal{L}_{\nu\phi} &+ \mathcal{L}_{nu-M} = \\ &= M \left(\nu_R \gamma^0 \nu_R + \text{h.c.}\right) + m_D \left(\bar{\nu}_R \nu_L + \bar{\nu}_L \nu_R\right). \end{aligned} \tag{6.36}$$

The problem of diagonalising the mass matrix obtained from (6.36) is discussed in section 13.3 of [1]. We summarise the results in the limit $M >> m_D$.

The eigenvectors of the mass matrix are two Majorana fields[2], ν' and ν'', with mass m' and $m'' >> m'$, respectively. The light neutrino coincides approximately with the neutrino emitted in β decay together with the electron:

$$\frac{1 - \gamma_5}{2} \nu' \simeq \nu_L; \quad m' \simeq \frac{m_D^2}{M} \tag{6.37}$$

while the heavy neutrino coincides approximately with ν_R.

$$\frac{1 + \gamma_5}{2} \nu'' \simeq \nu_R; \quad m'' \simeq M. \tag{6.38}$$

Since there are no restrictions on M, (6.37) allows a suppression of the mass of the light neutrino even for values of m_D of the order of the natural scale of the electroweak masses, fixed by the value of η in (6.22).

---

[2] i.e they are real fields $(\nu')^\dagger = \nu'$, $(\nu'')^\dagger = \nu''$, cf. chapter 13, [1].

CHAPTER 7

DETERMINATION OF THE NEUTRAL LEPTON CURRENT

CONTENTS

7.1   CROSS SECTIONS OF THE PROCESSES $\nu_\mu - e$ and $\bar{\nu}_\mu - e$

The current J_μ^Z in (4.32) is the sum of contributions from several leptons. Therefore in the product which appears in (4.34) there should be non-diagonal terms, for example terms which couple the muon neutrinos with the electron. In this section we study the cross section for elastic scattering processes of muon neutrinos and antineutrinos on atomic electrons:

$$\nu_\mu + e^- \rightarrow \nu_\mu + e^-$$
$$\bar{\nu}_\mu + e^- \rightarrow \bar{\nu}_\mu + e^- \tag{7.1}$$

measured with beams of high energy neutrinos, from a few GeV to several hundred GeV. In both cases, we can neglect the mass of the leptons, as we will do systematically in this chapter.

The processes (7.1) represent the simplest reactions predicted by the electroweak theory. They were discovered at the CERN PS in 1973, with the *Gargamelle* bubble chamber. High energy neutrino processes were studied at FermiLab, at the Tevatron, and at CERN, at the SPS, during the 1970s. The results from these experiments were of crucial importance for the development of the theory of elementary particles (for an extensive discussion of the historical development and the theory of neutrinos, cf. [14]).

The relevant terms of the Lagrangian (4.34) are:

$$\mathcal{L}^{(cn\,\nu_\mu e)} = \frac{G_F}{\sqrt{2}}[\bar{\nu}_\mu\gamma_\lambda(1-\gamma_5)\nu_\mu] \cdot [g_L\bar{e}\gamma^\lambda(1-\gamma_5)e + g_R\bar{e}\gamma^\lambda(1+\gamma_5)e];$$

$$g_L = -\frac{1}{2} + \sin^2\theta; \quad g_R = \sin^2\theta. \tag{7.2}$$

As an alternative, we can use the parametrisation in terms of axial and vector currents:

$$\mathcal{L}^{(cn\,\nu_\mu e)} = \frac{G_F}{\sqrt{2}}[\bar{\nu}_\mu\gamma_\lambda(1-\gamma_5)\nu_\mu] \cdot [\bar{e}\gamma^\lambda(g_V + g_A\gamma_5)e];$$

$$g_V = g_R + g_L = -\frac{1}{2} + 2\sin^2\theta;$$

$$g_A = g_R - g_L = \frac{1}{2}. \tag{7.3}$$

Reaction $\nu_\mu - e$.   The S-matrix element is written (we denote the 4-momenta with the symbols of the particles):

$$< \nu', e'|S^{(1)}|\nu, e > = (2\pi)^4\delta^{(4)}(\nu' + e' - \nu - e) \cdot \frac{G_F}{\sqrt{2}}(\Pi_{i,f}\sqrt{\frac{m}{EV}})\,\mathcal{M};$$

$$\mathcal{M} = \bar{u}(\nu')\gamma_\lambda(1-\gamma_5)u(\nu)\cdot$$
$$\cdot\,[g_L\bar{u}(e')\gamma^\lambda(1-\gamma_5)u(e) + g_R\bar{u}(e')\gamma^\lambda(1+\gamma_5)u(e)]. \tag{7.4}$$

The differential cross section in the laboratory frame of reference is:

$$d\sigma = (2\pi)^4\delta^{(4)}(\nu' + e' - \nu - e) \cdot \frac{d^3\nu'}{(2\pi)^3}\frac{d^3e'}{(2\pi)^3}\frac{G_F^2}{2}(\Pi_{i,f}\frac{m}{E})\left(\frac{1}{2}\sum_{all\ spin}|\mathcal{M}|^2\right). \tag{7.5}$$

The factor $\frac{1}{2}$ is due to the average over the initial spins; there are only two states to be averaged, the spin states of the electron, since as the neutrino originates from the decay of a pion it is in a pure helicity $\frac{1}{2}$ state.

We put:

$$m_e^2 m_\nu^2 \sum_{all\ spin}|\mathcal{M}|^2 = N^{\mu\nu}E_{\mu\nu}; \tag{7.6}$$

$$N^{\mu\nu} = \frac{1}{4}Tr\left[\slashed{\nu}\gamma^\mu(1-\gamma_5)\slashed{\nu}'\gamma^\nu(1-\gamma_5)\right] = \frac{1}{2}Tr\left[\slashed{\nu}\gamma^\mu\slashed{\nu}'\gamma^\nu(1-\gamma_5)\right] =$$
$$= 2\left[(\nu')^\mu\nu^\nu + (\nu')^\nu\nu^\mu - g^{\mu\nu}(\nu'\nu) + i\eta\epsilon^{\mu\alpha\nu\beta}\nu'_\alpha\nu_\beta\right]; \tag{7.7}$$

$$E^{\mu\nu} = \frac{1}{4}\left\{g_L^2Tr[\slashed{e}\gamma^\mu(1-\gamma_5)\slashed{e}'\gamma^\nu(1-\gamma_5)] + g_R^2Tr[\slashed{e}\gamma^\mu(1+\gamma_5)\slashed{e}'\gamma^\nu(1+\gamma_5)]\right\}$$
$$= 2\left\{g_L^2[(e')^\mu e^\nu + (e')^\nu e^\mu - g^{\mu\nu}(e'e) + i\eta\epsilon^{\mu\alpha\nu\beta}e'_\alpha e_\beta] + g_R^2(\eta \to -\eta)\right\}. \tag{7.8}$$

η is a sign which it is not worth the trouble to determine (but, however, cf. [1]); it is needed only to follow the relative signs between the neutrino terms and those of the electron. We find:

$$N^{\mu\nu}E_{\mu\nu} = 16[g_L^2(\nu e)(\nu' e') + g_R^2(\nu e')(\nu' e)] \tag{7.9}$$

from which:

$$d\sigma = (2\pi)^4 \delta^{(4)}(\nu' + e' - \nu - e) \cdot \frac{d^3\nu'}{(2\pi)^3} \frac{d^3e'}{(2\pi)^3} \cdot$$
$$\cdot 4(G_F^2 m_e) \frac{E_\nu}{E_{e'} E_{\nu'}} [g_L^2 + g_R^2 \left(\frac{(\nu' e)}{(\nu e)}\right)^2]. \tag{7.10}$$

The phase space integration is carried out in the standard way:

- the integral over the momentum of the outgoing neutrino eliminates the three-dimensional δ-function and sets $\nu' = \nu - e'$;

- the integration over the cosine of the polar angle, $\cos\theta$, of the final electron is carried out using the energy δ-function, whose argument is:

$$f(\cos\theta) = m_e + E_\nu - E_{e'} - \sqrt{E_\nu^2 + E_{e'}^2 - 2E_\nu E_{e'} \cos\theta} \tag{7.11}$$

and therefore introduces a factor:

$$\left|\frac{\partial f}{\partial \cos\theta}\right|^{-1} = \frac{E_{\nu'}}{E_\nu E_{e'}}; \tag{7.12}$$

- the integration over the azimuthal angle of the outgoing electron introduces a factor 2π.

In conclusion, we find:

$$d\sigma = \frac{dE_{e'}}{E_\nu} \frac{G_F^2 2m_e E_\nu}{\pi} [g_L^2 + g_R^2 \left(\frac{(\nu' e)}{(\nu e)}\right)^2]. \tag{7.13}$$

It is conventional to introduce the variable $0 \le y \le 1$:

$$y = \frac{E_{\nu'}}{E_\nu} = 1 + \frac{E_{e'}}{E_\nu} \tag{7.14}$$

and finally write:

$$\frac{d\sigma^{\nu,nc}}{dy} = \frac{G_F^2 s}{\pi} [g_L^2 + g_R^2(1-y)^2];$$

$$\sigma^{\nu,nc} = \frac{G_F^2 s}{\pi} [g_L^2 + \frac{1}{3} g_R^2]; \tag{7.15}$$

in terms of the variable s (the square of the centre of mass energy):

$$s = (e + \nu)^2 \simeq 2(e\nu) = 2m_e E_\nu. \tag{7.16}$$

The scale of the cross section is determined by the combination:

$$\frac{G_F^2 2m_e E_\nu}{\pi} = 1.72 \cdot 10^{-41} \text{ cm}^2 \frac{E_\nu}{1 \text{ GeV}}. \tag{7.17}$$

Reaction $\bar{\nu}_\mu - e$. We go from neutrino to antineutrino by simply exchanging $\nu \leftrightarrow \nu'$ in (7.7) or changing $\eta \to -\eta$ in (7.8). Finally we obtain the antineutrino cross section by simply changing $g_L \leftrightarrow g_R$ in (7.15):

$$\frac{d\sigma^{\bar{\nu},nc}}{dy} = \frac{G_F^2 s}{\pi}[g_L^2(1-y)^2 + g_R^2];$$

$$\sigma^{\bar{\nu},nc} = \frac{G_F^2 s}{\pi}[\frac{1}{3}g_L^2 + g_R^2]. \tag{7.18}$$

Charged current reactions. With a similar calculation the charged current cross section is obtained.

$$\nu_\mu + e \to \mu^- + \nu_e; \tag{7.19}$$

$$\frac{d\sigma^{\nu,cc}}{dy} = \frac{G_F^2 s}{\pi}. \tag{7.20}$$

The values of the neutral current chiral couplings are obtained from measurements of the ratios [14]:

$$R_\nu = \frac{\sigma^{\nu,nc}}{\sigma^{\nu,cc}} = g_L^2 + \frac{1}{3}g_R^2;$$

$$R_{\bar{\nu}} = \frac{\sigma^{\bar{\nu},nc}}{\sigma^{\nu,cc}} = \frac{1}{3}g_L^2 + g_R^2. \tag{7.21}$$

The ratios in (7.21) determine four combinations of g_L and g_R (the intersections of two ellipses) which differ by their signs.

The relative sign can be fixed by the cross section for $\nu_e - e$ processes at energies of the order of MeV, which is sensitive to terms of order $(m_e/E_\nu)g_L g_R$. The absolute sign is found from the forward-backward asymmetry in the process $e^+ e^- \to \mu^+ \mu^-$ (see later) which depends on the products $e g_L$ and $e g_R$.

In conclusion, from the neutrino data plus the information on the sign, it is found that:

$$g_A = 0.525 \pm 0.032;$$

$$g_V = -0.036 \pm 0.018 . \tag{7.22}$$

Although the precision of (7.22) was subsequently improved by measurements of the intermediate Z boson, neutrino reactions had a crucial importance because they gave a prediction of the mass of the intermediate bosons for the first time.

From (7.22) and (7.3) we find:

$$\sin^2 \theta \simeq 0.24 \tag{7.23}$$

and therefore ($e^2 = 4\pi\alpha \simeq 4\pi/137$):

$$M_W^2 = \frac{\pi\alpha}{\sqrt{2}G_F \sin\theta^2} \simeq (77.5 \text{ GeV})^2; \quad M_Z = \frac{M_W}{\cos\theta} \simeq 89.1 \text{ GeV}. \tag{7.24}$$

Note. The relation $\bar{e}_R\gamma^\mu e_L = 0$ implies that the axial and vector currents *conserve the particle helicities*. In the centre of mass of reaction (7.1), the helicities of the initial and final particles are distributed as in Figure 7.1, for the scattering of a neutrino by a L or R electron. In the second case, the initial component of the angular momentum in the direction of the electron is $(J_z)_{in} = +1$ while, if the scattering angle is 180°, $(J_z)_{fin} = -1$. Therefore the amplitude must vanish for $\cos\theta^* = -1$. Explicitly, we have:

$$1 - y = \frac{(\nu'e)}{(\nu e)} = \frac{1 + \cos\theta^*}{2}. \tag{7.25}$$

The factor $(1-y)^2$ which multiplies g_R^2 ensures the cross section satisfies the selection rule. Conversely, for the scattering of a neutrino on an L electron, the component of the angular momentum in the direction of the electron is $J_z = 0$ and the scattering amplitude need not have any zero.

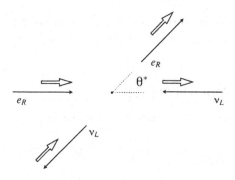

Figure 7.1 Conservation of helicity in $\nu_L - e_R$ scattering. For $\theta^* = 180°$ the angular momentum component along the line of flight is not conserved and the amplitude must vanish.

7.2 FUNDAMENTAL CONSTANTS

The values of the W and Z masses in (7.24) are in moderate agreement with the most recent experimental values:

$$(M_W)_{exp} = (80.425 \pm 0.03)\ \text{GeV}; \quad (M_Z)_{exp} = (91.1876 \pm 0.0021)\ \text{GeV}. \tag{7.26}$$

The differences are due to higher order corrections to the equations which relate the value of the masses to the fundamental constants α, G_F and $\sin^2\theta$. To remedy the fact that the error on the value of $\sin^2\theta$ from the cross sections for neutrino processes is not negligible, today it is preferred to use the value of

M_Z as a third fundamental constant and *to define* the value of $\sin^2\theta$ starting from equation (4.35), setting $G^{(nc)} = G_F$, or:

$$\sin^2(2\theta) = \frac{2\pi\sqrt{2}\alpha}{G_F M_Z^2} \tag{7.27}$$

To minimise the radiative corrections, it is furthermore preferred to insert the value of the fine structure constant defined at the Z mass rather than at zero energy, which is derived from atomic measurements. This results in a value of $\alpha(M_Z) \simeq 1/128$ (see later, section 8.4), with which it is found that:

$$\sin^2 2\theta(M_Z) = \frac{2\pi\sqrt{2}\alpha(M_Z)}{G_F M_Z^2} = 0.717 \rightarrow \sin^2\theta(M_Z) = 0.231 \tag{7.28}$$

from which:

$$M_W = \sqrt{\frac{\pi\alpha(M_Z)}{\sqrt{2}\sin^2\theta(M_Z)G_F}} = 79.8 \text{ GeV}. \tag{7.29}$$

The residual radiative corrections move further the predicted W mass and comparison with the experimental value represents a test of the theory to higher order.

7.3   LEPTONIC WIDTH OF THE Z

From (4.31) and (4.32) we find, to first order in g:

$$< f|S^{(2)}|Z > = \frac{1}{\sqrt{2M_Z V}}\epsilon^\mu \frac{g^2}{4\cos^2\theta} < f|J_Z^\mu(0)|0 > (2\pi)^4\delta^{(4)}(P-P_f). \tag{7.30}$$

Proceeding as usual, we obtain the formula for the width:

$$\Gamma = \frac{1}{3}\frac{1}{2M_Z}\frac{g^2}{4\cos^2\theta}\sum_f (2\pi)^4\delta^{(4)}(P - P_f)\cdot$$

$$\cdot < 0|J_Z^\nu(0)|f >< f|J_Z^\mu(0)|0 > (-g_{\mu\nu} + \frac{P_\mu P_\nu}{M_Z^2}). \tag{7.31}$$

If we neglect the masses of the charged leptons, the Z current is conserved and the projector on the states can be simplified in $-g_{\mu\nu}$. With simple calculations, and using the value of $\sin^2\theta$ given in (7.28), we find:

$$\Gamma(Z \rightarrow f + \bar{f}) = \Gamma_0(4g_L^2 + 4g_R^2);$$

$$\Gamma_0 = \frac{G_F}{12\sqrt{2}\pi}M_Z^3 \simeq 0.167 \text{ GeV}. \tag{7.32}$$

with

$$(4g_L^2 + 4g_R^2) = 1 \qquad (f = \nu_e, \nu_\mu, \nu_\tau);$$
$$(4g_L^2 + 4g_R^2) = (1 - 2\sin^2\theta)^2 + 4\sin^4\theta \simeq 0.502 \qquad (f = e, \mu, \tau). \tag{7.33}$$

The extension to the width of hadronic channels and the comparison with experimental data are considered later, in section 10.3.

7.4 RELATIVISTIC BREIT–WIGNER FORMULA

The propagator of a neutral particle (like the Z^0) must give the amplitude to find at time t the particle created at time 0 $(t > 0)$. For an unstable particle with lifetime τ and width $\Gamma = 1/\tau$, we expect that, as well as the factor $e^{(-i\omega t)}$, the amplitude contains a decreasing real exponential, which represents the *non-decay amplitude* up to time t:

$$A(t > 0) \sim e^{-i\omega(\mathbf{p})t} e^{-\frac{M}{\omega(\mathbf{p})}\frac{\Gamma}{2}t} \quad (t > 0). \tag{7.34}$$

(The factor $1/\gamma = M/\omega$ contains the relativistic time dilation of the lifetime of a particle moving with velocity $v = |\mathbf{p}|/\omega$.)

For times $t < 0$, the amplitude to observe the particle at 0, if it was produced at time t, must contain the factor:

$$A(t < 0) \sim e^{i\omega(\mathbf{p})t} e^{+\frac{M}{\omega(\mathbf{p})}\frac{\Gamma}{2}t} \quad (t < 0). \tag{7.35}$$

We recall that $A(t)$ is obtained by integration in the complex p^0 plane along the Feynman path [1]. More precisely, the exponents are the values of $-ip^0$ at the poles of the Fourier transform of the propagator. We see therefore that in the case of unstable particles the two poles must be found at:

$$\omega^+ = \omega(\mathbf{p}) - i\frac{\Gamma}{2\gamma} \Leftrightarrow (\omega^+)^2 \simeq \omega^2 - iM\Gamma; \tag{7.36}$$

$$\omega^- = -\omega(\mathbf{p}) + i\frac{\Gamma}{2\gamma} \Leftrightarrow (\omega^-)^2 \simeq \omega^2 - iM\Gamma; \tag{7.37}$$

where we have approximated with $\Gamma << M$. In terms of the variable p^2, we obtain the correct position of the two poles if we substitute in the propagator:

$$p^2 - M^2 \to p^2 - M^2 + iM\Gamma. \tag{7.38}$$

Correspondingly, the Fourier transform of the unstable Z propagator (cf. (5.71)) is obtained from the propagator of a stable vector field according to:

$$\frac{-g_{\mu\nu} + k_\mu k_\nu/M^2}{k^2 - M^2 + i\epsilon} \to \frac{-g_{\mu\nu} + k_\mu k_\nu/M^2}{k^2 - M^2 + iM\Gamma}. \tag{7.39}$$

Equation (7.39) is known as the *relativistic Breit–Wigner* formula.

7.5 PHYSICS OF THE Z IN e^+e^- REACTIONS

We consider the processes:

$$e^+e^- \to f\bar{f}; \quad (f \neq e) \tag{7.40}$$

To second order of perturbation theory and for $f \neq e$, the amplitude of the process is the sum of two amplitudes, corresponding respectively to the exchange of a photon and of a Z, (Figure 7.2):

$$\mathcal{M} = \mathcal{M}_\gamma + \mathcal{M}_Z. \tag{7.41}$$

Correspondingly, the squared modulus, and therefore the cross section, is the sum of three terms, which we denote respectively as the resonant, interference and electromagnetic terms:

$$|\mathcal{M}|^2 = |\mathcal{M}_Z|^2 + 2\mathcal{R}e(\mathcal{M}_Z \mathcal{M}_\gamma^*) + |\mathcal{M}_\gamma|^2. \tag{7.42}$$

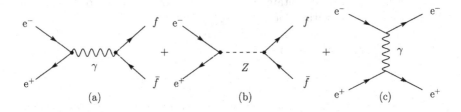

Figure 7.2 (a) and (b): Feynman diagrams for the annihilation $e^+ e^- \rightarrow f \bar{f}$, with $f \neq e$. For $f = e$ and small scattering angles it is necessary to take into account the diagram for Coulomb scattering, (c).

We set ourselves in the centre of mass of (7.40) and denote the total energy with E:

$$E = E_+ + E_- = 2E_+ . \tag{7.43}$$

Electromagnetic term. From the well known calculation of the cross section for the annihilation $e^+ e^- \rightarrow \mu^+ \mu^-$ [1] we obtain:

$$\sigma_\gamma(f\bar{f}) = \frac{4\pi\alpha Q_f^2}{3} \frac{1}{E^2}. \tag{7.44}$$

Resonant term. From the Lagrangian (4.31) we find[1]:

$$\sum_{fin} |\mathcal{M}_Z|^2 = (\frac{g^2}{4\cos^2\theta})^2 \sum_{fin} (< 0|J_\mu^Z(0)|e^+e^- > D_Z^{\mu\nu}(P) < fin|J_\nu^Z(0)|0 >)\cdot$$

$$\cdot (< e^+e^-|J_\lambda^Z(0)|0 > (D_Z^{\lambda\sigma}(P))^* < 0|J_\sigma^Z(0)|fin >)\cdot$$

$$\cdot (2\pi)^4\delta^{(4)}(p_+ + p_- - p_{fin}) =$$

$$= \frac{g^2}{4\cos^2\theta} |D(P)|^2\cdot$$

$$\cdot (< 0|J_\mu^Z(0)|e^+e^- >< e^+e^-|J_\lambda^Z(0)|0 >)\,\Pi^{\mu\nu}(P)\Pi^{\lambda\sigma}(P)\cdot$$

$$\cdot \frac{g^2}{4\cos^2\theta} \sum_{fin}(2\pi)^4\delta^{(4)}(p_+ + p_- - p_{fin})\cdot$$

$$\cdot < 0|J_\sigma^Z(0)|fin >< fin|J_\nu^Z(0)|0 > . \tag{7.45}$$

where we have set:

$$P = p_+ + p_-; \qquad |D(P)|^2 = \frac{1}{(P^2 - M_Z^2)^2 + M_Z^2\Gamma_Z^2};$$

$$\Pi^{\mu\nu}(P) = -g^{\mu\nu} + \frac{P^\mu P\nu}{M_Z^2}. \tag{7.46}$$

In the considerations which follow, we place ourselves in the region around the resonance, $P^2 \simeq M_Z^2$. In this region:

$$P_\mu\Pi^{\mu\nu}(P) \simeq 0. \tag{7.47}$$

The last line of (7.45) produces a tensor with two indices, σ and ν, which depend only on $P = p_+ + p_-$. This tensor can only be a combination of two possible tensors, $g_{\nu\sigma}$ and $P_\sigma P_\nu$, with the coefficients being functions of the invariant P^2. As a formula:

$$\frac{g^2}{4\cos^2\theta} \sum_{fin}(2\pi)^4\delta^{(4)}(P - p_{fin}) < 0|J_\sigma^Z(0)|fin >< fin|J_\nu^Z(0)|0 >=$$

$$= g_{\nu\sigma}\mathcal{A}(P^2) + P_\sigma P_\nu\mathcal{B}(P^2). \tag{7.48}$$

The term in $\mathcal{B}$ does not contribute as a result of the mass-shell condition (7.47), while $\mathcal{A}$ can be obtained by projecting with $\Pi^{\sigma\nu}$.

---

[1] From now on we omit the factors for the volume of normalisation, V, which cancel out in the cross section and/or width.

Noting that $\Pi^{\sigma\nu} g_{\sigma\nu} = 3$, we obtain:

$$\mathcal{A} = \frac{1}{3} \frac{g^2}{4\cos^2\theta} \Pi^{\sigma\nu}.$$
$$\cdot \sum_{fin} (2\pi)^4 \delta^{(4)}(P - p_{fin}) < 0|J_\sigma^Z(0)|fin >< fin|J_\nu^Z(0)|0 >=$$
$$= 2M_Z \Gamma_f;$$
$$\Gamma_f = \Gamma(Z \to fin). \tag{7.49}$$

The last step is obtained by comparing with equation (7.31) of the previous section.

Substituting in (7.45), we obtain:

$$\sum_{fin} |\mathcal{M}_Z|^2 = \frac{g^2}{4\cos^2\theta} |D(P)|^2 \cdot 2M_Z \Gamma_f.$$
$$\cdot g_{\sigma\nu} \Pi^{\mu\nu}(P)\Pi^{\lambda\sigma}(P) \ (< 0|J_\mu^Z(0)|e^+e^- >< e^+e^-|J_\lambda^Z(0)|0 >) =$$
$$= |D(P)|^2 \cdot 2M_Z \Gamma_f \cdot \Pi^{\mu\lambda}(\frac{g^2}{4\cos^2\theta} \ < 0|J_\mu^Z(0)|e^+e^- >< e^+e^-|J_\lambda^Z(0)|0 >). \tag{7.50}$$

If we sum over the spins of the electron and positron (as we must do anyway to obtain the non-polarised cross section) the last line of equation (7.50) can be expressed in terms of the partial width into e^+e^-. This is seen by considering the expression for $\Gamma(Z \to e^+e^-)$:

$$\Gamma(Z \to e^+e^-) = \Gamma_e =$$
$$= \frac{g^2}{4\cos^2\theta} \frac{1}{2M_Z} \int \frac{d^3p_+}{(2\pi)^3} \frac{d^3p_-}{(2\pi)^3} (2\pi)^4 \delta^{(4)}(P - p_+ + p_-) \cdot$$
$$\cdot \frac{1}{3} \Pi^{\mu\lambda}(P) \sum_{s\ in} < 0|J_\mu^Z(0)|e^+e^- >< e^+e^-|J_\lambda^Z(0)|0 > . \tag{7.51}$$

- We integrate p_- with the three-dimensional δ-function.

- We integrate the energy E_+ with the energy δ-function, obtaining a factor $\frac{1}{2}$ from the relation $\delta(2E_+ - M_Z) = \frac{1}{2}\delta(E_+ - M_Z/2)$.

- The result is invariant for rotations and therefore the integration of p_+ over solid angle gives 4π.

In conclusion, we find:

$$
\Gamma_e = \frac{1}{3}\frac{1}{2M_Z}\left(\frac{g^2}{4\cos^2\theta}\frac{M_Z^2}{8\pi}\Pi^{\mu\lambda}(P)\cdot\right.
$$

$$
\cdot\sum_{s\ in} < 0|J_\mu^Z(0)|e^+e^- >< e^+e^-|J_\lambda^Z(0)|0 >) =
$$

$$
= \frac{M_Z}{48\pi}\left(\frac{g^2}{4\cos^2\theta}\ \Pi^{\mu\lambda}(P)\sum_{s\ in} < 0|J_\mu^Z(0)|e^+e^- >< e^+e^-|J_\lambda^Z(0)|0 >\right).
$$

$$(7.52)$$

Finally substituting into (7.50) after having introduced the average over the initial spins, we find the simple result:

$$
\frac{1}{4}\sum_{s\ in}\sum_{fin}|\mathcal{M}_Z|^2 = \frac{24\pi\Gamma_e\Gamma_f}{(P^2-M_Z^2)^2+M_Z^2\Gamma_Z^2}. \tag{7.53}
$$

To obtain the resonant cross section we must simply divide by the flux factor, equal to $2v = 2$ in the centre of mass, and we obtain:

$$
\sigma_Z(e^+e^- \to f\bar{f}) = \frac{12\pi\Gamma_e\Gamma_f}{(P^2-M_Z^2)^2+M_Z^2\Gamma_Z^2}. \tag{7.54}
$$

Still in the centre of mass:

$$
P^2 = E^2 \tag{7.55}
$$

and the cross section, for $E \simeq M_Z$, is written:

$$
\sigma_Z = \frac{12\pi\Gamma_e\Gamma_f}{(E+M_Z)^2(E-M_Z)^2+M_Z^2\Gamma_Z^2} \simeq
$$

$$
\simeq \frac{12\pi}{M_Z^2}\frac{\Gamma_e\Gamma_f}{4(E-M_Z)^2+\Gamma_Z^2}. \tag{7.56}
$$

- The cross section at the peak does not contain any small parameter, and is therefore much larger than the electromagnetic cross section (7.44) which is of order α^2.

- For $E = M_Z \pm \frac{1}{2}\Gamma_Z$ the cross section decreases by a factor two compared to the peak value; Γ_Z can be obtained from the variation of the cross section as the *width at half height*.

- The exchange of the Z gives an excellent approximation to the total cross section around the Z pole also for $f = e$, except for events in which the final particles are very close to the initial direction, for which it is necessary to take into account the Coulomb scattering amplitude, diagram (c) in Figure 7.2.

The results obtained suggest a simple strategy to verify the $SU(2)_L \otimes U(1)_Y$ theory:

- The variation of cross section for any observable channel (e.g. $f = \mu$) allows accurate determination of M_Z and Γ_Z.

- The measurement of the peak cross section

$$\sigma_{peak} = \frac{12\pi}{M_Z^2} B_e B_f \qquad (7.57)$$

(B denotes the decay branching fraction) allows the determination of the branching fraction, thus of the width, of all the *visible* channels, e, μ, τ, hadrons.

From each of these measurements we obtain different determinations of $\sin^2 \theta$ which should be self-consistent among themselves and with the neutrino measurements.

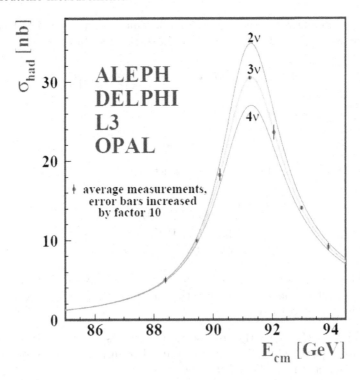

Figure 7.3 Electron–positron cross section for $E \simeq M_Z$. The theoretical curves are shown as a function of the total number of light *left-handed* or *right-handed* neutrinos. The data confirm the Standard Theory, which predicts three *left-handed* neutrinos ν_e, ν_μ, ν_τ, with remarkable precision. Figure from [15].

- We can determine by subtraction the width for the *invisible channels*: decays into particles which do not interact in the detectors, like neutrinos:

$$\Gamma_{inv} = \Gamma_Z - \Gamma_{vis} \tag{7.58}$$

and compare the invisible width with what is expected for neutrinos.

$$\Gamma_{inv} = N\Gamma_0. \tag{7.59}$$

Γ_0 is the width given in equation (7.32); N is the number of neutrino types with $V-A$ (left handed) coupling into which the Z can decay, in practice neutrinos with a mass less than about $M_Z/2$.

- As an alternative, we can sum all the visible channels in the cross section and compare with σ_Z in (7.56):

$$\sigma_{Z,vis}(E) = \frac{12\pi}{M_Z^2} \frac{\Gamma_e(\Gamma_Z - N\Gamma_0)}{4(E - M_Z)^2 + \Gamma_Z^2}. \tag{7.60}$$

The comparison is shown in Figure 7.3 and demonstrates excellent agreement with $N = 3$, in agreement with the known neutrinos, ν_e, ν_μ, ν_τ.

Differential and interference cross sections.

$$D_H D_J^* = \left(\sum_{spin} < 0|H_\mu|e^+ e^- >< e^+ e^-|J_\nu|0 > \right) \cdot$$

$$\cdot \left(\sum_{spin} < f\bar{f}|H_\mu|0 >< 0|J_\nu|f\bar{f} > \right) \quad H, J = J^{em}, J^Z; \quad (7.61)$$

direct terms: $H = J$, interference: $H \neq J$. Explicitly ($M = M_Z$, $\Gamma = \Gamma_Z$):

$$|D_J|^2 = \frac{1}{P^4} \quad (J = J^{em}); \quad |D_J|^2 = \frac{1}{(P^2 - M^2)^2 + M^2\Gamma^2} \quad (J = J^Z);$$

$$D_{J^{em}} D_{J^Z}^* + D_{J^{em}}^* D_{J^Z} = \frac{P^2 - M^2}{P^2[(P^2 - M^2) + M^2\Gamma^2]}. \tag{7.62}$$

We see that *the interference vanishes at the peak of the resonance.*

With easy steps, repeating what was already done, we find (in the centre of mass and in the limit of massless e and f, and all energies equal to $E = \frac{1}{2}M$; we denote the 4-momenta with the corresponding particle symbols):

$$E_{\mu\nu} = \left(\sum_{spin} < 0|H_\mu|e^+ e^- >< e^+ e^-|J_\nu|0 > \right) =$$

$$= \frac{1}{2E_e^2} Tr \{\slashed{e}\gamma_\mu \slashed{e}\gamma_\nu [g_L^e h_L^e (1 - \gamma_5) + g_R^e h_R^e (1 + \gamma_5)]\} \tag{7.63}$$

where $g^e_{L,R}$ and $h^e_{L,R}$ are the chiral constants for the electron of the currents J and H.

$$\left(\sum_{spin} <0|H_\mu|e^+e^- ><e^+e^-|J_\nu|0> \right)_{LL}$$

$$= g^e_L h^e_L \frac{2}{E^2_e} \left[e_\nu \bar{e}_\mu + e_\mu \bar{e}_\nu - g_{\nu\mu}(e\bar{e}) + i\eta\epsilon_{\nu\alpha\mu\beta}e^\alpha\bar{e}^\beta \right] ;$$

$$\left(\sum_{spin} <f\bar{f}|H_\mu|0><0|J_\nu|f\bar{f}> \right)_{LL}$$

$$= g^f_L h^f_L \frac{2}{E^2_e} \left[e_\nu \bar{e}_\mu + e_\mu \bar{e}_\nu - g_{\nu\mu}(e\bar{e}) + i\eta\epsilon_{\mu\alpha\nu\beta}e^\alpha\bar{e}^\beta \right] ; \quad (7.64)$$

where η is the factor ± 1 introduced in section 7.1 and that is similarly not necessary to specify. Note the exchange $(\nu \leftrightarrow \mu)$ between the two lines of equation (7.64).

Overall, we find:

$$\left(\sum_{spin} <0|H_\mu|e^+e^- ><e^+e^-|J_\nu|0> \right) \cdot \left(\sum_{spin} <f\bar{f}|H_\mu|0><0|J_\nu|f\bar{f}> \right)$$

$$= \frac{16}{E^4_e} \cdot \left[(e\bar{f})(\bar{e}f)(g^e_L h^e_L \cdot g^f_L h^f_L + g^e_R h^e_R \cdot g^f_R h^f_R) + \right.$$

$$\left. + (ef)(\bar{e}\bar{f})(g^e_L h^e_L \cdot g^f_R h^f_R + g^e_R h^e_R \cdot g^f_L h^f_L) \right]. \quad (7.65)$$

We define θ as the angle between the direction of the electron and that of the fermion f (therefore the angle of $\bar{f}$ is $\theta + \pi$). It is easily seen that:

$$(e\bar{f}) = (\bar{e}f) = E^2_e(1 + \cos\theta); \quad (ef) = (\bar{e}\bar{f}) = E^2_e(1 - \cos\theta) \quad (7.66)$$

from which:

$$\left(\sum_{spin} <0|H_\mu|e^+e^- ><e^+e^-|J_\nu|0> \right) \cdot \left(\sum_{spin} <f\bar{f}|H_\mu|0><0|J_\nu|f\bar{f}> \right)$$

$$= 16 \cdot [(1 + \cos^2\theta)(g^e_L h^e_L + g^e_R h^e_R)(g^f_L h^f_L + g^f_R h^f_R) +$$

$$+ 2\cos\theta(g^e_L h^e_L - g^e_R h^e_R)(g^f_L h^f_L - g^f_R h^f_R)]; \quad (7.67)$$

$$A^f_{FB} = \frac{\int d\sigma(\cos\theta > 0) - \int d\sigma(\cos\theta < 0)}{\int d\sigma(\cos\theta > 0) + \int d\sigma(\cos\theta < 0)}. \quad (7.68)$$

Problem. Show that if ϕ has spin $= 0$ the cross section for the reaction $e^+e^- \to \phi + \bar{\phi}$ is equal to $\frac{1}{4}$ of the cross section given in equation (7.44).

QUARKS IN THREE COLOURS

CONTENTS

In this chapter we describe the developments which brought us to our present picture (cf. chapter 1) according to which hadrons[1] are constituted of fundamental spin $\frac{1}{2}$ particles, the quarks, of several types known as *flavours*, each one in three varieties, called *colours*.

8.1　SAKATA MODEL

As a starting point we can take the model of Sakata mentioned in chapter 2, according to which the hadrons are bound states of particles from a triplet of fundamental constituents, identified as the baryons p, n, Λ and their antiparticles.

According to the Sakata model, the mesons are constituted of $S\bar{S}$ pairs, where S denotes the general element of the triplet (2.51). This structure gives rise to $3 \times 3 = 9$ pseudoscalar mesons and as many vector mesons, which perfectly reproduce the quantum numbers of the *nonets* of observed particles, Figures 2.1 and 2.2. For example (L denotes the orbital angular momentum

---

[1]We recall that this name denotes particles sensitive to strong interactions.

of the $S\bar{S}$ pair and s the total spin):

$$\mathbf{S\bar{S}} : \mathbf{L = 0, s = 0}:$$
$$\pi^+ = (p\bar{n}); \ K^+ = (p\bar{\Lambda}); \ etc.$$
$$\mathbf{S\bar{S}, \ L = 0, s = 1}:$$
$$\rho^+ = (p\bar{n}); \ K^{*+} = (p\bar{\Lambda}); \ etc.$$
$$(8.1)$$

The fundamental symmetry of the Sakata model is the $SU(3)$ group of unitary and unimodular transformations of the fundamental triplet, equation (2.51). The fundamental representation is denoted by 3, from its dimensionality, and the conjugate representation, according to which the antiparticles $\bar{S}$ transform, is denoted by $\bar{3}$. The states $S\bar{S}$ transform as the tensor product $3 \times \bar{3}$ which, as we will see, factorises into the regular representation, 8, and a singlet:

$$3 \otimes \bar{3} = 8 \oplus 1 . \qquad (8.2)$$

The problem of the Sakata model is with the baryons. The simplest hypothesis is that, as well as the triplet S, they should be classified as $SS\bar{S}$ states, Since Λ has strangeness $S(\Lambda) = -1$, the baryons which contain $\bar{\Lambda}$ should have strangeness $S = +1$. Instead, from Figures 2.3 and 2.4, it is clear that all the light baryons have negative strangeness.

8.2 EIGHTFOLD WAY AND QUARKS

The problem was resolved in two steps.

In the first, following Gell–Mann and Ne'eman [16], we consider the $SU(3)$ symmetry in a more abstract way, as a symmetry which acts on the fundamental degrees of freedom of the hadrons, *whatever they are*, along the lines indicated for isotopic spin in chapter 2.

If g is an element of $SU(3)$, all that we require to have an $SU(3)$ symmetry is that the observed particles transform according to representations $g \to U(g)$ of $SU(3)$. For example, for the mesons:

$$M \to U_M(g)M \qquad (8.3)$$

and we know that the representation $U_M = 8 \oplus 1$ reproduces the quantum numbers of the observed particles.

This point of view leaves us completely free to choose the representation according to which the baryons transform. And furthermore it has the advantage of being able to treat *all* the observed baryons on the same level.

Gell–Mann and Ne'eman proposed to choose the 8 representation for the spin $\frac{1}{2}$ baryons[2] and the 10 representation for the spin $\frac{3}{2}$ baryons.

---

[2]The name *Eightfold Way* given by Gell–Mann to his scheme originates here.

At the time of the proposal, the eight spin $\frac{1}{2}$ baryons were already known. If we reorganise the particles in Figure 2.3 as a function of $Y = S+1$ instead of S, we obtain exactly the same scheme as for the mesons, except for the absence of the singlet. The 10 representation takes into account all the spin $\frac{3}{2}$ particles observed at the time of the proposal, except the state with $Y = -2$ (or $S = -3$), cf. Figure 2.4. The observation, in 1964, of the Ω^- which completed the decuplet represented the crucial confirmation of the $SU(3)$ symmetry in the realisation of the Eightfold Way.

The second step, due to Gell–Mann[17] and Zweig[18], completed the scheme by identifying the fundamental degrees of freedom. If we assume that these degrees of freedom are associated with an $SU(3)$ triplet (as in the Sakata model), the observed particles should transform according to representations which are tensor products of 3 and $\bar{3}$ representations.

The fundamental observation is that, with spin $\frac{1}{2}$ constituents, we have:

$$\text{Mesons}: \ 3 \otimes \bar{3} = 8 \oplus 1; \tag{8.4}$$

$$\text{Baryons}: \ 3 \otimes 3 \otimes 3 = 1 \oplus 8 \oplus 8 \oplus 10 . \tag{8.5}$$

In the fundamental triplet, the generators are represented by the Gell–Mann matrices, introduced in chapter 2. If we return to the formulae (2.61) and (2.60), we see that the constituents must have exotic quantum numbers: $B = 1/3$ and $Q = +2/3, -1/3, -1/3$. Therefore the constituents are not among the observed hadronic particles, unlike what happens in the Sakata model and its predecessor, the model of Fermi and Yang. These particles, which represent an entirely new level of matter, were called *quarks* by Gell–Mann, both name[3] and concept having been extraordinarily successful.

Following Gell–Mann, the fundamental triplet is written as:

$$q = \begin{bmatrix} u \\ d \\ s \end{bmatrix} \tag{8.6}$$

a notation which underlines the quark quantum numbers, which unite isotopic spin ($u = up$ and $d = down$) and strangeness ($s = strange$). These quantum numbers, and those associated with other types of quark later discovered, have become known as *flavours*.

Tensors and representations of $SU(3)$. Many of the general results on the representations of a Lie group described in Appendix C can be reconstructed starting from the tensor representations, formed from tensor products of the fundamental representations of the group.

---

[3]The name originates from a phrase in the book *Finnegans Wake* by James Joyce: *"Three quarks for Muster Mark!"*.

In the case of $SU(2)$ the fundamental representation corresponds to spin $\frac{1}{2}$ and is constituted of complex vectors q_i, with $i = 1, 2$:

$$q'_i = U_{ij} q_j; \quad U^\dagger U = 1; \quad \text{Det } U = 1; \tag{8.7}$$

and the representations of $SU(2)$ are described by tensors $q_{ij\ldots l}$ with n *symmetric indices*, corresponding to spin$=\frac{1}{2}n$.

A first fundamental representation of $SU(3)$ is constructed, in analogy with the fundamental representation of $SU(2)$, by extending the number of indices from two to three. However, as we will see in a moment, the complex conjugate tensor transforms in a non-equivalent way. In $SU(3)$ there are two fundamental, non-equivalent representations, usually distinguished by the notation 3 (lower index) and $\bar{3}$ (upper index):

$$
\begin{aligned}
q'_a &= U_{ab} q_b, & \text{representation } (3); \\
(\bar{q}')^a &= U^*_{ab} (\bar{q})^b, & \text{representation } (\bar{3}); \\
U^\dagger U &= 1, \quad \text{Det } U = 1.
\end{aligned}
\tag{8.8}
$$

From these rules the invariant tensors and the invariant operations are constructed, to project the irreducible components of the representation formed from the tensor products of 3 and $\bar{3}$. The latter are characterised by tensors with n_1 upper indices and n_2 lower indices (the procedure is completely analogous to what was done for the representations of the Lorentz group in [1]) and the invariant operations are:

- symmetrisation or antisymmetrisation of two or more indices, either upper or lower,

- contraction, by multiplication with δ^a_b, of one upper and one lower index,

- contraction of two or three indices, either upper or lower, with the completely antisymmetric tensors ϵ_{abc} and ϵ^{abc}.

The invariance of the tensors ϵ_{abc} and ϵ^{abc} corresponds to the fact that the matrices U have determinant $= 1$, as in the case of the Lorentz group [1]. In the case of $SU(2)$, the antisymmetric tensor, ϵ_{ij}, allows to move from the fundamental representation to one which transforms like its conjugate representation. The fact that the invariant tensor now has three indices makes the 3 and $\bar{3}$ representations non-equivalent.

The representations which remain invariant under *all* the operations above are those characterised by tensors, $T^{a_1 a_2 \cdots a_{n_1}}_{b_1 b_2 \cdots b_{n_2}}$, with n_1 upper indices and n_2 lower indices, which are *symmetric in the upper indices and in the lower indices, and have zero trace*:

$$
\begin{aligned}
T^{a_1 a_2 \cdots a_{n_1}}_{b_1 b_2 \cdots b_{n_2}} &= T^{a_2 a_1 \cdots a_{n_1}}_{b_1 b_2 \cdots b_{n_2}}, \text{ etc.} \\
T^{a a_2 \cdots a_{n_1}}_{a b_2 \cdots b_{n_2}} &= 0.
\end{aligned}
\tag{8.9}
$$

It can be shown [70] that these tensors describe all the irreducible representations of $SU(3)$, which are therefore characterised by *two integer numbers*, n_1 and n_2, corresponding to the fact that $SU(3)$ has *two* commuting generators. We give a few simple examples of how the reduction of tensor products into irreducible components is carried out.

(i): $\mathbf{3} \otimes \bar{\mathbf{3}}$

$$v^a w_b = T_b^a = \hat{T}_b^a + \frac{1}{3}\delta_b^a(\delta_d^c T_c^d) = \hat{T}_b^a + \delta_b^a T \qquad (8.10)$$

where $\hat{T}_b^a$ has zero trace. The space of these zero trace tensors has dimensions $3 \cdot 3 - 1 = 8$, therefore:

$$3 \otimes \bar{3} = 8 \oplus 1; \qquad (8.11)$$

(cf. equation (8.4)).

(ii): $\mathbf{3} \otimes \mathbf{3}$

$$v_a w_b = T_{a,b} = T_{ab} + T_{[ab]}^A \qquad (8.12)$$

where T and T^A are, respectively, symmetric and antisymmetric. A symmetric tensor with indices from 1 to 3 has six independent components; an antisymmetric tensor has three of them and can be written as:

$$T_{[ab]}^A = \epsilon_{abc}\bar{q}^c. \qquad (8.13)$$

Therefore:

$$T_{a,b} = T_{ab} + \epsilon_{abc}\bar{q}^c, \qquad \text{or}:$$
$$3 \otimes 3 = 6 \oplus \bar{3}. \qquad (8.14)$$

(iii): $\mathbf{3} \otimes \mathbf{6}$ We use the relation

$$w_a T_{bc} = T_{abc} + \epsilon_{abd}T_c^d$$

where T_{abc} is symmetric in the three indices and

$$T_c^d = \epsilon^{abd}w_a T_{bc} \quad \text{from which} \quad T_d^d = \epsilon^{abd}w_a T_{bd} = 0$$

for the symmetry of T_{bd}. Therefore T_c^d is in the 8 representation. The tensor with three symmetric indices has ten components (see later) and, in conclusion:

$$6 \otimes 3 = 10 \oplus 8. \qquad (8.15)$$

(iv): $\mathbf{3} \otimes \mathbf{3} \otimes \mathbf{3}$

$$3 \otimes 3 \otimes 3 = (\bar{3} \oplus 6) \otimes 3 = 1 \oplus 8 \oplus 8 \oplus 10 \qquad (8.16)$$

(cf. equation (8.5)).

The counting of the components of the representation with three symmetric indices can be done as follows. Given the symmetry of T_{abc} we can restrict ourselves to the components with $a \geq b \geq c$. There is only one component with 333, two components with two 3s, and three components with only one 3: 322, 321, 311; in total six components in which 3 appears. The other cases have only indices which take values 1 or 2, i.e. make a symmetric $SU(2)$ tensor of rank 3, which represents a spin $\frac{3}{2}$ tensor with four components. Therefore, in total, T_{abc} has ten components.

The index 3 is associated with the strange quark (S = −1). The argument just made therefore shows the decomposition of the 10 from $SU(3)$ into $SU(2) \otimes U(1)_S$. Going back, we have: a multiplet with isospin $\frac{3}{2}$ and zero strangeness, an isospin 1 and strangeness −1, an isospin $\frac{1}{2}$ and strangeness −2 and an isospin 0 and strangeness −3, Figure 2.4.

Problem: decomposition of 6 and $\bar{3}$ into $SU(2) \otimes U(1)_S$. If we represent the multiplets with isotopic spin I and strangeness S with the notation (I, S), we have

$$6 = (1,0) \oplus (1/2, -1) \oplus (0, -2);$$
$$\bar{3} = (0,1) \oplus (1/2, 0), \quad \text{or} \quad (\bar{s}) \oplus (\bar{d}, \bar{u}).$$

8.3 QUARK COLOURS

Towards the end of the 1960s, it could be claimed that the quark model elegantly accounted for the spectra of the lightest mesons and baryons, and of the other resonances discovered up until then.

However doubts persisted about the true nature of quarks, by some considered merely a mathematical device, useful to describe the spectrum of the hadrons but lacking an actual connection to reality. The doubts were in large part due to the problem of the symmetry of the quark wave function of the baryons and furthermore to the failure of all attempts made up to then to observe quarks in high energy collisions or in nature, as stable particles[4] left over from the Big Bang at the origin of the universe.

On the first point, it is a fact that, to describe the overall structure of the spin and charge of the baryons, it is necessary that the state of three quarks should be *completely symmetric* for exchange of the quarks, in conflict with

---

[4]Because of the fractional value of the electric charge, the lightest quark should be absolutely stable.

the spin-statistics relation which requires that spin $\frac{1}{2}$ particles obey Fermi–Dirac statistics and therefore should have a completely antisymmetric wave function.

The problem appears immediately by considering the Δ^{++} resonance in the $S_z = +3/2$ state. In the quark model, this state corresponds to

$$\Delta^{++}(S_z = 3/2) = \int \left[u^\uparrow(x_1) u^\uparrow(x_2) u^\uparrow(x_3) \right] f(x_1, x_2, x_3) d^3 x_1 d^3 x_2 d^3 x_3. \quad (8.17)$$

where $f(x_1, x_2, x_3)$ is a suitable wave function in the coordinate space of the quarks. In the configuration with orbital angular momentum $L = 0$, $f(x_1, x_2, x_3)$ is completely symmetric and we obtain a completely symmetric state under the exchange of quarks.

Ignoring these doubts, there were attempts to describe the primary strong interactions between quarks as due to the exchange of an electrically neutral vector particle, similar to the photon, which was given the name of *gluon*, from 'glue', after its role in binding the quarks inside hadrons.

The problem of the symmetry of the baryon wave function, for spin and flavour, finds a natural solution if we assume that each quark of a given flavour possesses an additional quantum number which takes three values. There are now numerous states which can represent the $\Delta^{++}(S_z = 3/2)$, of the form:

$$(u_i^\uparrow u_j^\uparrow u_k^\uparrow) \quad (8.18)$$

and it is possible to satisfy the Pauli principle, if we assume that the $\Delta^{++}(S_z = 3/2)$, and in general all baryons, should be in a *completely antisymmetric* state in the new quantum numbers.

In 1965, Han and Nambu [19] gave an elegant formulation of this hypothesis, introducing an $SU(3)'$ symmetry which operates on the indices $i, j, k = 1, \cdots, 3$. The desired state is obtained by projecting with the antisymmetric tensor:

$$\Delta^{++}(S_z = 3/2) =$$
$$= \int \left[u_i^\uparrow(x_1) u_j^\uparrow(x_2) u_k^\uparrow(x_3) \right] \epsilon^{ijk} f(x_1, x_2, x_3) d^3 x_1 d^3 x_2 d^3 x_3. \quad (8.19)$$

As we just saw in the case of flavour $SU(3)$, the antisymmetric tensor and the previous formula identify the state of the baryons as the *singlet state* of $SU(3)'$. An analogous state can be created with the mesons and we can hypothesise, at least provisionally, that:

- *the observed hadron states of mesons and baryons, are singlets of the $SU(3)'$ symmetry.*

In the formulation of Han and Nambu, the different coloured quarks also have a different distribution of electric charge. This allows to attribute integer electric charges to them, in contrast to what happens in the model of Gell–Mann and Zweig. In the following equation (8.20) we summarise the electric charge values attributed to the u, d, and s quarks for the three values of the $SU(3)'$ index.

$$
Q = \begin{pmatrix} +1 & +1 & 0 \\ 0 & 0 & -1 \\ 0 & 0 & -1 \end{pmatrix} \begin{matrix} u \\ d \\ s \end{matrix}
$$
$$
\begin{matrix} 1 & 2 & 3 \end{matrix} \tag{8.20}
$$

However, towards the end of the 1970s, an alternative solution was considered, in which the quarks have fractional electric charge and the new quantum number is completely independent of flavour. Each quark of a given flavour is present in three variants which we can characterise with three different 'colours': red, green and blue. The values of electric charge attributed to the u, d and s quarks are shown in equation (8.21)

$$
Q = \begin{pmatrix} +2/3 & +2/3 & +2/3 \\ -1/3 & -1/3 & -1/3 \\ -1/3 & -1/3 & -1/3 \end{pmatrix} \begin{matrix} u \\ d \\ s \end{matrix}
$$
$$
\begin{matrix} 1 & 2 & 3 \end{matrix} \tag{8.21}
$$

In the colour picture, the weak interactions are based on the flavour currents which commute with colour and the formulation of a Yang–Mills theory of only the electroweak interactions, like the Glashow–Weinberg–Salam theory, is logically possible.

An important result from 1972 due to Bouchiat, Iliopoulos and Meyer marked a point in favour of quarks of three colours. In the electroweak theory of quarks and leptons, based on the gauge group $SU(2)_L \otimes U(1)$, the anomalies which could prevent the renormalisability cancel between the quarks and leptons, if the quarks have fractional charge and exist in three colours [20].

8.4 QUANTUM CHROMODYNAMICS

We can reformulate the theory of the gluon, considering a Yang–Mills theory based on the colour symmetry, which we denote as $SU(3)_c$ to distinguish it from the flavour symmetry, $SU(3)_f$. As mediators of the primary strong interactions, there are now eight vector fields, as many as the $SU(3)_c$ generators. They are all electrically neutral fields and still known by the name *gluons*.

The Lagrangian which describes quarks and gluons in interactions is known as *quantum chromodynamics* (QCD) and is written extending the general form given in (3.35) to the case of $SU(3)_c$:

$$\mathcal{L}_{QCD} = \sum_f \bar{q}_f (iD_\mu \gamma^\mu + m_f) q_f - \frac{1}{4} \sum_A G^A_{\mu\nu} (G^A)^{\mu\nu};$$

$$iD_\mu = i\partial_\mu - g_S \frac{\lambda^A}{2} g^A_\mu;$$

$$(G^A)^{\mu\nu} = \partial_\nu g^A_\mu - \partial_\mu g^A_\nu + g_S f^{ABC} g^B_\mu g^C_\nu; \qquad (8.22)$$

where:

- q_f is the quark field of flavour f (the colour index is implied), g^A_μ the field of the gluon;

- g_S is the coupling constant, λ^A the Gell–Mann matrices in the space of colour and f^{ABC} the $SU(3)_c$ structure constant.

The study of QCD is beyond the scope of the present volume (for an introduction to the argument, see the third volume of the present series [3]). We limit ourselves to the illustration of a few results.

Global flavour symmetries.   We rewrite the fermion term of (8.22) in terms of the left- and right-handed fields:

$$\mathcal{L}_{QCD,q} = \sum_f \bar{q}_{fL} \, iD_\mu \gamma^\mu \, q_{fL} + \sum_f \bar{q}_{fR} \, iD_\mu \gamma^\mu q_{fR} +$$

$$+ \sum_f m_f \left(\bar{q}_{fL} \, q_{fR} + \bar{q}_{fR} \, q_{fL} \right). \qquad (8.23)$$

The kinetic term in the first line of (8.23) is invariant under the chiral transformations generated by the pairs of unitary matrices $N_f \times N_f$:

$$q_{fL} \to q'_{fL} = U_{f'f} q_{fL}; \qquad q_{fR} \to q'_{fR} = V_{f'f} q_{fR} \qquad (8.24)$$

that is, invariant under the chiral symmetry group:

$$G(N_f) = U(N_f)_L \otimes U(N_f)_R$$
$$= U(1)_V \otimes U(1)_A \otimes SU(N_f)_L \otimes SU(N_f)_R \qquad (8.25)$$

which clearly also leaves the gluon term in (8.22) invariant.

The mass term is invariant under the first abelian factor, which corresponds to conservation of overall baryon number, but in general reduces $G(N_f)$ to a symmetry group of lesser dimensions, according to the values of the masses of the different flavours. As we will see in later chapters, a good approximation is to consider the masses of the quarks c, b and t as generally large and different from one another, and to consider three limiting cases:

Eightfold way :

$SU(N_f)_L \otimes SU(N_f)_R \rightarrow SU(3)$ $(m_s = m_u = m_d);$

Isotopic spin :

$SU(N_f)_L \otimes SU(N_f)_R \rightarrow SU(2)$ $(m_s >> m_u = m_d);$

Chiral $SU(2)$:

$SU(N_f)_L \otimes SU(N_f)_R \rightarrow SU(2)_L \otimes SU(2)_R \; m_s \neq 0;$ $(m_u = m_d = 0).$

$$(8.26)$$

The first case corresponds to the $SU(3)_f$ flavour symmetry considered in the previous section, the second to isotopic spin symmetry, chapter 2, and the third case will be considered in chapter 11.

The general consequence of (8.22) is that, in any case, the symmetries in flavour space are violated *only* by the quark mass terms, and not by the gluon interactions, an outcome which has been successfully tested in several specific cases.

Asymptotic freedom. The dependence of the amplitudes of a renormalisable field theory on the momentum scale applicable in a given process, for example $e^+e^- \rightarrow \mu^+\mu^-$, has been studied by many authors, in the asymptotic limit in which the momentum scale tends to infinity.

In this limit, it can be shown that the dominant higher order corrections can be summarised by a dependence of the fundamental constants of the theory on the momentum scale. The behaviour of the effective, or *running*, coupling constant as a function of the momentum scale, q^2, is determined by the so-called renormalisation group equations, established in the 1950s by Gell–Mann and Low.

The decisive step forward for the acceptance of QCD as a fundamental theory of nuclear interactions was made by Gross and Wilczek [21] and, independently, by Politzer [22], in 1973.

The explicit calculation of the Gell–Mann and Low equation for a non-abelian gauge theory, like QCD, showed that the effective constant tends asymptotically to zero as q^2 increases:

$$\alpha_S(q^2) = \frac{g_S^2(q^2)}{4\pi} \approx \frac{C}{Log(q^2)} \rightarrow 0 \qquad (q^2 \rightarrow \infty). \qquad (8.27)$$

The result remains valid while the number of quark flavours in the theory does not exceed 16, amply in excess of the number of observed flavours, which is 6.

The result in (8.27), completely unexpected at the time, explains the scaling relations in the cross-sections of electron-proton and neutrino-proton scattering processes at high momentum transfer observed at the end of the 1960s and in the early part of the 1970s.

The scaling law, first recognised by Bjorken, was interpreted by Feynman as due to the scattering of the electron or neutrino on elementary constituents of the proton (*partons*) which react *as though* they are free of forces. This phenomenon was given the suggestive name of 'asymptotic freedom' of the strong interactions.

Feynman himself had pointed out that about 50% of the momentum of the proton is carried by electrically neutral particles, which in QCD are naturally associated with gluons. The experimental investigation of electron and neutrino scattering at high momentum transfer (*deep inelastic scattering*) had led to the conclusion that the proton (and every other hadron) participated in high momentum transfer processes *as if* they were constituted of an incoherent set of gluons and quarks of different flavours, each one characterised by a function, $P_i(x)$, which gives the probability of finding within the proton a parton of a given type $i = u, d, s, \cdots, g$, with a fraction x of the proton momentum. The functions $P_i(x)$ are known as *structure functions*, or *parton distribution functions*, of the proton (or other hadrons, where relevant).

The logarithmic approach to asymptotic freedom causes the scaling relations to be affected by logarithmic corrections, which have been calculated precisely by many authors [23] and compared with experimental data at increasingly high energies. The extraordinary agreement between theoretical predictions and the experimental observations today constitutes solid evidence in favour of QCD as the theory of the strong interactions (see for example the survey in [24]).

Infrared slavery. In the opposite direction to that of (8.27), when the momentum scale is reduced, the coupling constant grows strongly, leaving the perturbative region at a critical value defined by Λ_{QCD}, with:

$$\Lambda_{QCD} = 200 - 300 \text{ MeV}. \tag{8.28}$$

For momenta below Λ_{QCD} a regime is encountered which today we do not control theoretically, but in which the powerful forces related to colour charges prevent the coloured objects, quarks and gluons, from being individually observed; only colour singlets should have finite energy . In this case one speaks of *confinement* of quarks and gluons, or also of *infrared slavery*[5].

The possibility that the same theory could explain the apparent freedom of quarks and gluons in high momentum transfer reactions and their non-observation in a free state in nature underlies the proposal to take QCD as the fundamental theory of strong interactions [25, 26].

Comparison with QED. The high momentum behaviour of a theory, like QED, based on an abelian gauge group is strikingly different.

The effective constant of QED grows logarithmically with the momentum

---

[5]In obvious contrast to the asymptotic freedom in the ultraviolet.

scale, a result due to Landau. The rate of increase is linked to the probability that a photon of very large momentum, q^2, materialises as a lepton-antilepton or quark-antiquark pair. This probability, in its turn, can be determined starting from the measurement of the cross-section for electron-positron annihilation into lepton or quark pairs, in the approximation in which the annihilation proceeds via the exchange of a virtual photon (see [1]).

From data taken at LEP, it is possible to estimate that the fine structure constant changes from the value obtained with spectroscopic data, $\alpha \simeq 1/137$, which can be considered as associated with momentum scales of order of the electron mass, to the value at the mass of the Z^0, see [27]:

$$\alpha(M_Z^2) \simeq \frac{1}{128}. \tag{8.29}$$

QCD coupling constant at the M_Z scale. Similarly to what happens for the fine structure constant, in electroweak calculations it is often convenient to use the value of the QCD coupling constant at the mass scale of the Z^0. In the following we will use the value [27]:

$$\alpha_S(M_Z^2) = 0.118 . \tag{8.30}$$

THEORY OF QUARK MIXING

CONTENTS

The idea of mixing between quarks as an effect of symmetry breaking was introduced by Cabibbo in 1963 [28], in the context of a quark theory with three flavours: *up*, *down* and *strange*. In this picture, it is natural to classify the left-handed fields *up* and *down* as an electroweak doublet, and the other fields as singlets, according to the scheme:

$$\begin{pmatrix} u \\ d \end{pmatrix}_L ; \quad u_R; \ d_R; \ s_L; \ s_R. \tag{9.1}$$

9.1 THE CABIBBO THEORY

Cabibbo observed that symmetry breaking can give rise to a mixing between d_L and s_L, giving an expression for the weak current of the form:

$$J_\mu^1 + i J_\mu^2 = \bar{u}_L \gamma_\mu \left(\cos\theta d_L + \sin\theta s_L \right). \tag{9.2}$$

With a single electroweak parameter, the Cabibbo angle θ, and a few phenomenological parameters related to strong interactions, excellent agreement for β decays of baryons and mesons, with and without strangeness, is obtained.

Introducing the quark triplet:

$$q_L = \begin{pmatrix} u \\ d \\ s \end{pmatrix}_L, \tag{9.3}$$

the previous equation is written:

$$J_\mu^1 + i J_\mu^2 = \bar{q}_L \mathcal{C} \gamma_\mu q_L =$$

$$= \bar{q}_L \begin{pmatrix} 0 & \cos\theta & \sin\theta \\ 0 & 0 & 0 \\ 0 & 0 & 0 \end{pmatrix} \gamma_\mu q_L. \tag{9.4}$$

The Cabibbo theory, however, cannot be incorporated in the electroweak theory of Glashow–Weinberg–Salam, since equation (9.4) leads to a *neutral electroweak current which changes the strangeness*, in contrast with experimental data. The neutral current actually involves the commutator of the matrix $\mathcal{C}$ with its adjoint, which is not diagonal:

$$[\mathcal{C}, \mathcal{C}^\dagger] = \begin{pmatrix} 1 & 0 & 0 \\ 0 & -\cos^2\theta & -\sin\theta\cos\theta \\ 0 & -\sin\theta\cos\theta & -\sin^2\theta \end{pmatrix}. \tag{9.5}$$

Terms of the form $\bar{d}_L \gamma_\mu s_L$ in the neutral current produce the decay $K_L \to \mu^+\mu^-$ with a decay rate of the order of that of $K^+ \to \mu^+\nu$. For this reason, in their original papers Glashow, Weinberg and Salam restricted their considerations to the electroweak interactions of the electron-neutrino doublet.

9.2 THE GLASHOW–ILIOPOULOS–MAIANI MECHANISM

The difficulty was overcome in 1970 by Glashow, Iliopoulos and Maiani [29], with the introduction of a fourth, *charm*, quark with the same electroweak quantum numbers as the *up* quark. In the presence of the *charm* quark, we can define the s_L quark as an electroweak doublet, according to the scheme:

$$\begin{pmatrix} u \\ d \end{pmatrix}_L ; \begin{pmatrix} c \\ s \end{pmatrix}_L ; \quad u_R; \ d_R; \ c_R; \ s_R. \tag{9.6}$$

The matrix $\mathcal{C}$ now becomes a 4×4 matrix and the weak current takes the form (quarks are ordered as u, c, d and s, and each element is a 2×2 block):

$$J_\mu^1 + i J_\mu^2 = \bar{q}_L \begin{pmatrix} 0 & U_{GIM} \\ 0 & 0 \end{pmatrix} \gamma_\mu q_L ;$$

$$U_{GIM} = \begin{pmatrix} \cos\theta & \sin\theta \\ -\sin\theta & \cos\theta \end{pmatrix}. \tag{9.7}$$

The neutral current, owing to the unitarity of U_{GIM}, is completely diagonal:

$$J_\mu^3 = \bar{q}_L \left[\mathcal{C}, \mathcal{C}^\dagger \right] \gamma_\mu q_L =$$

$$= \bar{q}_L \begin{pmatrix} U_{GIM} U_{GIM}^\dagger & 0 \\ 0 & -U_{GIM}^\dagger U_{GIM} \end{pmatrix} \gamma_\mu q_L =$$

$$= \bar{q}_L \begin{pmatrix} 1 & 0 \\ 0 & -1 \end{pmatrix} \gamma_\mu q_L. \tag{9.8}$$

The mass of the charm quark. The GIM paper [29] considers a theory with a single charged intermediate boson coupled to the current $J_\mu^1 + iJ_\mu^2$ given by (9.4). In this scheme, the neutral current (9.5) is not coupled and therefore there are no forbidden transitions to lowest order of perturbation theory.

However, to the next higher order the transition $K_L \to W^+W^- \to \mu^+ + \mu^-$ can occur, which gives rise to the process (10.17); see Figure 9.1. In a pioneering paper, Ioffe and Shabalin [30] pointed out that, in the Cabibbo theory, the diagram of Figure 9.1 with the *up* quark only gives rise to a *divergent* result, equal to:

$$\mathcal{M}(K_L \to \mu^+\mu^-) = constant \times \sin\theta \cos\theta \times G(G\Lambda^2) \tag{9.9}$$

where Λ is a cut-off energy beyond which the theory loses validity, and *constant* is a numerical factor. The surprising fact is that the comparison with the observed value of the neutral current transition gives a cut-off value of order of a few GeV, an energy already reached at that time and at which there had been no observation of any special anomaly.

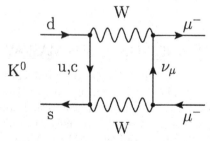

Figure 9.1 Feynman diagram for the process $K^0 = (d\bar{s}) \to \mu^+\mu^-$ according to GIM.

A more detailed analysis shows that the coefficient of (9.9) is actually proportional to the non-diagonal matrix element of the neutral current in (9.5) which, eliminated at lowest order, reappears in the higher order corrections. The observation made in [29] is that the amplitude in which the *charm* quark is exchanged, in the diagram of Figure 9.1, interferes negatively with the amplitude due to the *up* quark considered by Ioffe and Shabalin so as to produce a convergent result (the *GIM mechanism*) of order:

$$\mathcal{M}(K_L \to \mu^+ \mu^-) = constant \times \sin\theta\cos\theta \times G\left[G(m_c^2 - m_u^2)\right]. \quad (9.10)$$

The result (9.10), compared with experimental data (10.17) allowed to estimate for the first time the *order of magnitude of the charm quark mass* (assuming $m_c \gg m_u$). The most constrained estimate arises from the K_L–K_S mass difference, which is also connected to the exchange of two W bosons. It gives:

$$m_c = 2 - 3 \text{ GeV}, \quad (9.11)$$

a value consistent with the mass of the lightest *charm* meson, $D = (c\bar{q})$ ($q = u,\ d$):

$$M_{D^+} = M(c\bar{d}) = 1869.3 \text{ MeV}; \quad M_{D^0} = M(c\bar{u}) = 1864.5 \text{ MeV}. \quad (9.12)$$

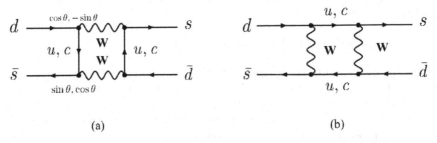

(a)   (b)

Figure 9.2   Feynman diagrams for the transitions $K^0 \to \bar{K}^0$ according to GIM.

9.3   *CP* VIOLATION AND KOBAYASHI–MASKAWA THEORY

In 1973 the situation concerning electroweak unification had completely changed compared to 1970. The work of Veltman and 't–Hooft had shown that the Weinberg–Salam theory was renormalisable and therefore an excellent candidate for a *fundamental* theory of the electromagnetic and weak interactions. The existence of the *charm* quark and the GIM mechanism, necessary for the suppression of strangeness changing neutral current processes, was already accepted by the majority of theoretical physicists.

However, it still remained to be understood what was the source of *CP* violation, observed in the decay of K mesons since 1964. A fundamental electroweak theory should include *CP* violation, the most natural way being the presence of one or more complex phases in the mixing matrix. However, it is easy to show [29] that, in a theory with two doublets, U_{GIM} can always be made real, by exploiting the arbitrariness in the phase of the quark fields.

The simplest way of obtaining this result is to start from the most general expression for the Cabibbo current (the first row of the U_{GIM} matrix):

$$J_\mu^C = \bar{u}_L\gamma_\mu \left(e^{i\alpha}\cos\theta d_L + e^{i\beta}\sin\theta s_L\right) \tag{9.13}$$

and to notice that we can reabsorb the two phases into the definition of d_L and s_L. At this point, the second row of U_{GIM} is fixed, at least to an overall phase, from the condition of being orthonormal to the first row. We thus obtain:

$$\begin{aligned}
J_\mu^1 + iJ_\mu^2 &= \bar{u}_L\gamma_\mu \left(\cos\theta d_L + \sin\theta s_L\right) \\
&\quad + \bar{c}_L\gamma_\mu e^{i\phi}\left(-\sin\theta d_L + \cos\theta s_L\right)
\end{aligned} \tag{9.14}$$

but now we can absorb the phase $e^{i\phi}$ into the field $\bar{c}_L$ and obtain the purely real form given in (9.7).

The (justified!) question which was posed by Kobayashi and Maskawa [31] was: how many quark fields are needed to retain at least one phase which cannot be eliminated? The example to study is that of N left-handed doublets and N right-handed singlets. In this situation, it is easy to see that $N = 3$ actually provides the minimum case, with a single uneliminable phase. On this basis Kobayashi and Maskawa proposed that at least one other fundamental doublet should exist, beyond the two doublets identified in (9.6), so that CP violation could be included in the Weinberg-Salam theory.

Proof.   We start from the fact that an $N \times N$ unitary matrix, U, is characterised by N^2 real parameters. This can be seen from the fact that we can always write:

$$U = e^{iH}$$

with H Hermitian, and H has:

N real parameters on the diagonal,

$2 \times \dfrac{N(N-1)}{2} = N^2 - N$ real parameters in elements above the diagonal,

in total $N_{un} = N^2$ real parameters $\qquad\qquad$ (9.15)

(the elements below the diagonal are not independent since they are complex conjugates of the elements above it).

On the other hand, an orthogonal $N \times N$ matrix is described by N_{ort} real parameters, with:

$$N_{ort} = \frac{N(N-1)}{2}$$

and therefore in the unitary matrix there are N_f complex phases, where:

$$N_f = N_{un} - N_{ort} = N^2 - \frac{N(N-1)}{2} = \frac{N(N+1)}{2}. \tag{9.16}$$

However, we have a total of $2N$ complex fields in the left-handed doublets and we can dispense with $2N - 1$ phases by reabsorbing them in these fields (-1 comes from the fact that current remains unchanged by a change of phase common to all fields of the *up* and *down* type). The total number of irreducible phases is N_{irr}:

$$N_{irr} = \frac{N^2 + N - 4N + 2}{2} = \frac{N^2 - 3N + 2}{2} = \frac{(N - 1)(N - 2)}{2}. \qquad (9.17)$$

We recover the known results for $N = 1$ and 2 and, remarkably, we obtain $N_{irr} = 1$ in the case of three doublets.

During 1976, several new results appeared:

- The first direct evidence for particles with *charm*.

- Evidence for a heavy lepton, the τ, which increased to three the number of lepton families.

- Evidence for a new quark of the *down* type, the b or *beauty* quark.

In subsequent years, various experiments obtained evidence that the τ and b belonged to an electroweak doublet, thus confirming the picture illustrated in chapter 1, which was later completed with the direct observation of the *top* quark in 1994 and the ν_τ neutrino in 1999.

In 2001, experiments carried out at KEP (Japan) and SLAC (USA) demonstrated excellent agreement between the CP violation observed in decays of b mesons as calculated starting from the phase predicted by Kobayashi and Maskawa.

These discoveries amply justify the now universal convention of associating the names of Cabibbo, Kobayashi and Maskawa with the U_{CKM} matrix which characterises quark mixing in the theory with three families.

ELECTROWEAK INTERACTIONS OF QUARKS

CONTENTS

In this chapter we consider the incorporation of subnuclear matter into the Weinberg–Salam theory.

Hadrons, the particles sensitive to strong interactions, are described by states composed of quarks and antiquarks. In nature, three families, or generations, of quarks are observed with the same electroweak quantum numbers, according to the scheme:

$$
\begin{pmatrix} u \\ d \end{pmatrix}_L ; \quad u_R, \; d_R;
$$

$$
\begin{pmatrix} c \\ s \end{pmatrix}_L ; \quad c_R, \; s_R;
$$

$$
\begin{pmatrix} t \\ b \end{pmatrix}_L ; \quad t_R, \; b_R. \tag{10.1}
$$

There are three varieties of each quark field distinguished by a colour quantum number with which the strong interactions of the quarks are associated (e.g., $u \rightarrow u_\alpha$, $\alpha = 1, 2, 3$; cf. section 8.3).

In analogy with the electron and the corresponding neutrino, we have attributed weak isotopic spin $\frac{1}{2}$ and 0 to the *left-handed* and *right-handed* fields, respectively. In the presence of breaking of the electroweak symmetry, however, it is necessary to distinguish the fields which have definite isospin in the previous equations from the fields which create and destroy the physical quarks.

In the following, we denote with u_{ph}, d_{ph} etc. the physical fields, meaning that the fields without a suffix are those with a definite weak isospin.

In parallel with the three quark families, the known leptons are organised into three families of fields, devoid of colour and with **definite weak isospin**.

$$\begin{pmatrix} \nu_e \\ e \end{pmatrix}_L \; ; \; e_R;$$

$$\begin{pmatrix} \nu_\mu \\ \mu \end{pmatrix}_L \; ; \; \mu_R;$$

$$\begin{pmatrix} \nu_\tau \\ \tau \end{pmatrix}_L \; ; \; \tau_R. \tag{10.2}$$

The previous assignments determine the interactions of quarks and leptons with the electroweak gauge fields, intermediate bosons and the photon. The corresponding quantum numbers have been confirmed experimentally in various processes, in particular in the study of decays of the vector bosons, Z and $W^\pm$; see section 10.3.

For completeness and for later reference we give in Table 10.1 the values of the fundamental fermion masses and the relevant couplings to the neutral intermediate boson, Z^0.

Table 10.1 Masses of the charged fermions of the three generations, in GeV, and their coupling to Z^0.

	I Gen.	II Gen.	III Gen.	$4(g_L^2 + g_R^2)$
up-like	u: 0.0044	c: 1.42	t: 172.5	$(1 - \frac{4}{3}\sin^2\theta_W)^2 + \frac{16}{9}\sin^4\theta_W = 0.571$
down-like	d: 0.0077	s: 0.150	b: 4.49	$(-1 + \frac{2}{3}\sin^2\theta_W)^2 + \frac{4}{9}\sin^4\theta_W = 0.737$
charged lepton	e: 0.000511	μ: 0.1016	τ: 1.777	$(-1 + 2\sin^2\theta_W)^2 + 4\sin^4\theta_W = 0.502$

For the lightest quarks, *up*, *down* and *strange*, we have given the masses obtained in chapter 11 starting from chiral symmetry violation. For the other quarks, the value is deduced from the mass of the corresponding $q\bar{q}$ meson, the so-called *on-shell* mass, at the momentum scale equal to the value of the mass itself.

The chiral couplings are derived from equation (4.32) with appropriate values of the (fractional) electric charge of the quarks, and $\sin^2 \theta_W = \sin^2 \theta_W(M_Z^2)$,

10.1   ORIGIN OF QUARK MASS AND MIXING

For every pair of families, we can construct two invariants analogous to those in (6.29) and (6.33). Explicitly (the family indices i and j are fixed; the repeated indices of weak isospin, $a, b = 1, 2$, are summed):

$$\mathcal{L}_{ij} = g_{ij}^D \bar{Q}_i \phi D_{R,j} + g_{ij}^U \epsilon_{ab} \bar{U}_{R,i} Q_j^a \phi^b + h.c. \qquad (i, j = 1, 2, 3);$$

$$\mathcal{L}_{qH} = \sum_{ij} \mathcal{L}_{ij}. \qquad (10.3)$$

Q_i represents the general left-handed doublet in (10.1) while $U_{R,i}$ and $D_{R,i}$ denote the right-handed fields of up type (u, c, t) and $down$ type (d, s, b). The coupling constants g_{ij}^D and g_{ij}^U can take any, eventually complex, value, since we require hermiticity of the Lagrangian only; complex values of the coupling constants lead to violations of the CP and T symmetries, leaving the CPT symmetry exact, cf. [1].

If we substitute the vacuum expectation value of the Higgs field (cf. (6.16)) into (10.3) we obtain the quark mass Lagrangian:

$$\mathcal{L}_{qm} = \bar{D}_L M^d D_R + \bar{U}_R M^u U_L + h.c.;$$
$$(M^d)_{ij} = g_{ij}^D \eta; \quad (M^u)_{ij} = g_{ij}^U \eta. \qquad (10.4)$$

$M^{d,u}$ are matrices in the space of the families, in general complex and non-diagonal. This introduces a difference between the electroweak basis, identified by the fields given in (10.1), and the physical basis identified by fields which diagonalise the mass matrices and are associated with physical particles. To characterise the physical basis, we use the decomposition of the $M^{d,u}$ matrices illustrated in Appendix B, eq. (B.12), for which we can write:

$$M^u = W^\dagger m_u Z; \qquad M^d = U^\dagger m_d V, \qquad (10.5)$$

with $m_{u,d}$ diagonal and positive matrices and W, Z, U, V unitary matrices. The mass Lagrangian is therefore written as:

$$\mathcal{L}_{qm} = \bar{D}_L U^\dagger m_d V D_R + \bar{U}_R W^\dagger m_u Z U_L + h.c. \qquad (10.6)$$

We now note that the transformations:

$$U_R \to W U_R; \qquad D_R \to V D_R \qquad (10.7)$$

correspond to a redefinition of the electroweak singlets which we can realise respecting the electroweak symmetry. Similarly, we can absorb the matrix Z in a redefinition of the left-handed doublet:

$$Q = \begin{pmatrix} U_L \\ D_L \end{pmatrix} \to ZQ \qquad (10.8)$$

and thus obtain:

$$\mathcal{L}_{qm} = \bar{D}_L Z U^\dagger m_d D_R + \bar{U}_R m_u U_L + h.c. =$$
$$= \bar{D}_L U_{CKM} m_d D_R + \bar{U}_R m_u U_L + h.c. \qquad (10.9)$$

The fields that appear in (10.9) still have definite weak isospin but, while the fields of type U diagonalise the mass, in the new basis there is still a misalignment for the D-type fields. We therefore set:

$$(D_{ph})_L = U^\dagger_{CKM} D_L; \text{ or :}$$

$$D_L = \begin{pmatrix} d \\ s \\ b \end{pmatrix}_L = U_{CKM} \quad (D_{ph})_L = U_{CKM} \begin{pmatrix} d_{ph} \\ s_{ph} \\ b_{ph} \end{pmatrix}_L ; (10.10)$$

$$D_{ph} = (D_{ph})_L + D_R. \qquad (10.11)$$

In terms of the physical fields, the mass Lagrangian is now completely diagonal:

$$\mathcal{L}_{qm} = \bar{D}_L Z U^\dagger m_d D_R + \bar{U}_R m_u U_L + h.c. =$$
$$= \bar{D}_L U_{CKM} m_d D_R + \bar{U}_R m_u U_L + h.c. =$$
$$= (\bar{D}_{ph})_L m_d D_R + (\bar{U}_{ph})_R m_u (U_{ph})_L + h.c. =$$
$$= \bar{D}_{ph} m_d D_{ph} + \bar{U}_{ph} m_u U_{ph}. \qquad (10.12)$$

As we will see shortly, the matrix U_{CKM} is actually the Cabibbo–Kobayashi–Maskawa matrix introduced in the previous chapter, which determines the weak interactions of the neutral quark current.

The misalignment represented by U_{CKM} breaks the electroweak symmetry. The lowering operator for weak isospin applied to $U = U_{ph}$ produces $D \neq D_{ph}$, which is a superposition of physical fields associated with particles with different masses and therefore not an eigenstate of the Hamiltonian. Physically, this effect is transmitted to the current coupled to the charged intermediate boson. The current, associated with the operator which raises the weak isospin, changes each *down*-type quark, in the electroweak basis, into the corresponding *up*-type field:

$$J_\mu^{(+)} = J_\mu^1 + i J_\mu^2 = \bar{U}_L \gamma_\mu D_L = (\bar{U}_{ph})_L \gamma_\mu U_{CKM} (D_{ph})_L =$$

$$= \frac{1}{2} \begin{pmatrix} \bar{u} & \bar{c} & \bar{t} \end{pmatrix} \gamma_\mu (-\gamma_5) U_{CKM} \begin{pmatrix} d \\ s \\ d \end{pmatrix} \qquad (10.13)$$

where, from now on, we omit the ph suffix from the quark fields, for notational simplicity.

The diagonal elements of U_{CKM}, which are found experimentally to dominate, produce the inter-family decays, $d \to u$, $c \to s$ and $t \to b$. The non-diagonal elements violate the conservation of family type. We therefore conclude that, from the mixing effect induced by the breaking of $SU(2)_L \otimes U(1)_Y$

- *processes mediated by the charged currents violate quark flavour conservation.*

Beyond this, the matrix elements of U_{CKM} have a cosmological significance by permitting decays of the s and b quarks, the lightest particles of their respective doublets, which would otherwise be totally stable. Their presence explains why ordinary matter is constituted only of quarks of u and d type.

To complete the picture of the interactions of the quarks with the $SU(2)_L \otimes U(1)_Y$ vector fields, we must consider the coupling of the Z boson, which comes about through the neutral current (cf. equation (4.31)):

$$J_\mu^{(Z)} = J_\mu^3 - \sin^2 \theta J_\mu^{e.m.} \tag{10.14}$$

where $J^{e.m.}$ represents the electromagnetic current and J^3 the third component of the weak isospin current. If we adopt the notation used in (10.8) to denote the set of left-handed doublets, we can write (the matrix T^+ is composed of 3×3 blocks):

$$J_\mu^1 + iJ_\mu^2 = \bar{Q}\gamma_\mu T^+ Q = \bar{Q}\gamma_\mu \begin{pmatrix} 0 & U_{CKM} \\ 0 & 0 \end{pmatrix} Q \tag{10.15}$$

and, correspondingly:

$$J_\mu^3 = \bar{Q}\gamma_\mu \left[T^+, T^- \right] Q =$$
$$= \bar{Q}\gamma_\mu \begin{pmatrix} U_{CKM}U_{CKM}^\dagger & 0 \\ 0 & -U_{CKM}^\dagger U_{CKM} \end{pmatrix} Q$$
$$= \bar{Q}\gamma_\mu \begin{pmatrix} 1 & 0 \\ 0 & -1 \end{pmatrix} Q. \tag{10.16}$$

Given that the electromagnetic current is clearly diagonal in the quarks, the neutral current implied by (10.14) and by the mixing scheme (10.11) is diagonal in the families. In contrast to charged current processes

- *processes mediated by the neutral current conserve quark flavour in lowest order of the electroweak interactions.*

The first experimental evidence for this selection rule comes from the decays of K mesons, for which it is found that:

$$B(K^+ \to \mu^+ \nu) = (63.55 \pm 0.11) \cdot 10^{-2}$$
$$B(K_L^0 \to \mu^+ \mu^-) = (6.84 \pm 0.11) \cdot 10^{-9} \tag{10.17}$$

and from the $K^0 \bar{K}^0$ mass difference. Processes of this type may arise in higher orders, as described by the diagrams of Figures 10.1 and 9.2. The amplitudes are suppressed by the GIM mechanism, or by its extension to the theory with six flavours, so as to be effectively of order G_F^2, as observed.

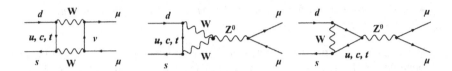

Figure 10.1 The relevant Feynman diagrams for the exchange of vector bosons in the process $K^0 = (d\bar{s}) \to \mu^+\mu^-$ in the electroweak theory. To these one should add the diagram with the exchange of the Higgs boson.

10.2   THE CABIBBO–KOBAYASHI–MASKAWA MATRIX

Over the years, the parameterisation of the U_{CKM} matrix due to Wolfenstein [32](cf. [33] for a recent summary) has proven particularly useful. This parameterisation starts from the observation that the matrix elements are of order unity on the diagonal and become increasingly small for transitions between quarks of higher generations. The order of magnitude of the transitions is given by $\lambda = \sin\theta_C$, corrected by two real parameters: one real parameter, A, for $b \to c$, and one parameter $A^2(\rho - i\eta)$, for $b \to u$. The CP-violating phase is due to $\eta \neq 0$.

Once the basic transitions are fixed, the other elements of the matrix are determined by its unitarity. The unitarity relations are solved in a power series in λ up to the desired precision. We give here the parameterisation introduced in [34] (see also [35]) in which terms up to order λ^5 are retained.

$$U_{CKM} =$$

$$\begin{pmatrix} 1 - \frac{1}{2}\lambda^2 - \frac{1}{8}\lambda^4 & \lambda & A\lambda^3(\rho - i\eta) \\ -\lambda + \frac{1}{2}A^2\lambda^5[1 - 2(\rho + i\eta)] & 1 - \frac{1}{2}\lambda^2 - \frac{1}{8}\lambda^4(1 + 4A^2) & A\lambda^2 \\ A\lambda^3[1 - (1 - \frac{1}{2}\lambda^2)(\rho + i\eta)] & -A\lambda^2 + \frac{1}{2}A\lambda^4[1 - 2(\rho + i\eta)] & 1 - \frac{1}{2}A^2\lambda^2 \end{pmatrix}$$

$$\tag{10.18}$$

In many cases, it is sufficient to use Wolfenstein's original parameterisation [32], valid to order λ^3:

$$U_{CKM} = \begin{pmatrix} 1 - \frac{1}{2}\lambda^2 & \lambda & A\lambda^3(\rho - i\eta) \\ -\lambda & 1 - \frac{1}{2}\lambda^2 & A\lambda^2 \\ A\lambda^3[1 - (\rho + i\eta)] & -A\lambda^2 & 1 \end{pmatrix}. \tag{10.19}$$

The values of the parameters of the CKM matrix are obtained by a simultaneous fit to various data on electroweak interactions of particles with different flavours; cf. [33] for an extensive discussion. In the same reference the best fit values are given:

$$\lambda = 0.2253 \pm 0.0007, \quad A = 0.808^{+0.022}_{-0.015},$$
$$\bar{\rho} = 0.132^{+0.022}_{-0.014}, \quad \bar{\eta} = 0.341 \pm 0.013, \tag{10.20}$$

where $\bar{\rho}$ and $\bar{\eta}$ are roughly equal to ρ and η, respectively, and connected to each other by the relation:

$$\rho + i\eta = \frac{(\bar{\rho} + i\bar{\eta})\sqrt{1 - A^2\lambda^4}}{\sqrt{1 - \lambda^2}[1 - A^2\lambda^4(\bar{\rho} + i\bar{\eta})]}. \tag{10.21}$$

These relations ensure that $\bar{\rho} + i\bar{\eta} = -(U_{ud}U_{ub}^*)/(U_{cd}U_{cb}^*)$ is independent of the phase convention and that the CKM matrix, written in terms of $\bar{\rho}$ and $\bar{\eta}$, is unitary to all orders of λ.

Figure 10.2 shows the constraints arising from the different physical measurements and demonstrates the quality of the fit in (10.20).

10.3 Z AND W DECAY WIDTHS

The combination of the chiral constants in Table 10.1 determines the width of the Z^0 into different quark-antiquark pairs. Repeating the considerations made for the leptonic width and again in the limit of massless fermions in the final state, from equation (7.32) we obtain:

$$\Gamma(Z \to q + \bar{q}) = \Gamma_q = \Gamma_0 \cdot 3 \cdot (4g_L^2 + 4g_R^2);$$
$$\Gamma_0 = \frac{G_F}{12\sqrt{2}\pi}M_Z^3 \simeq 0.167 \text{ GeV} \tag{10.22}$$

where the factor 3 represents the multiplicity of quark colours.

Summing over the available channels (five active quark flavours and three of leptons) we find:

$$\Gamma(Z \to \text{hadrons}) = 2\Gamma_u + 3\Gamma_d = 1.67 \text{ GeV} \quad (\text{exp}: 1.744 \pm 0.002 \text{ GeV});$$
$$\Gamma(Z \to l^+l^-) = 0.249 \text{ GeV} \quad (\text{exp}: 0.2519 \pm 0.0003 \text{ GeV});$$
$$\Gamma(Z \to \text{invisible}) = 0.497 \text{ GeV} \quad (\text{exp}: 0.499 \pm 0.002 \text{ GeV}). \tag{10.23}$$

The width into l^+l^- represents the sum of the e, μ and τ channels. We include in parentheses the observed values [27].

The agreement for the lepton channels is excellent. For example, dividing the experimental value of the invisible width by the width into a single neutrino we find $N_\nu = 3.01$.

In the case of hadrons, the equations of the renormalisation group, section 8.4, predict a further correction factor with which:

$$3 \to 3(1 + \frac{\alpha_S}{\pi}) \tag{10.24}$$

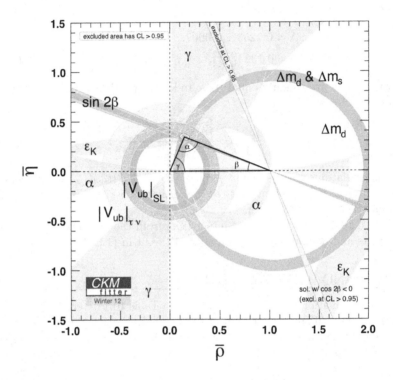

Figure 10.2 CKM fit. The coordinates of the unitarity vertex are determined by the intersection of geometric loci corresponding to different observables $\epsilon_K, \Delta m$, etc. and fix the values of the CKM parameters ρ and η. Figure from CKM Fitter. For more details see [33].

in (10.22). Using equation (8.30), the correction corresponds to:

$$1.67 \text{ GeV} \rightarrow 1.73 \text{ GeV} \tag{10.25}$$

in excellent agreement with the hadronic width.

We leave to the reader the task of deriving the analogous formulae for W decays and just quote the results (see [27] for the comparison with data):

$$\Gamma(W \rightarrow e^+\nu_e) = \frac{G_F}{6\sqrt{2}\pi} M_W^3 = 0.227 \text{ GeV};$$

$$\Gamma(W \rightarrow u\bar{d} + u\bar{s} + u\bar{b}) = \Gamma_q =$$

$$= \frac{G_F}{6\sqrt{2}\pi} M_W^3 \cdot 3 \sum_{i=1,3} |(U_{CKM})_{1i}|^2 = 0.681 \text{ GeV}. \tag{10.26}$$

Summing over the two allowed hadronic processes ($u\bar{d} + \cdots$ and $c\bar{d} + \cdots$) and over the three lepton channels, we find:

$$\Gamma(W \to \text{hadrons}) = 1.36 \text{ GeV} \quad (1.41 \pm 0.03 \text{ GeV});$$
$$\Gamma(W \to \text{leptons}) = 0.681 \text{ GeV} \quad (0.686 \pm 0.013 \text{ GeV}). \quad (10.27)$$

With the extra QCD correction given in (10.24) we obtain, for the hadronic width, the prediction:

$$1.36 \text{ GeV} \to 1.41 \text{ GeV} \quad (10.28)$$

in excellent agreement with the data.

10.4   THE STANDARD THEORY LAGRANGIAN

To conclude this chapter, we can assemble the complete Lagrangian of the *Standard Theory*, the theory which describes electroweak and strong interactions of quarks, gluons, leptons, electroweak bosons and Higgs fields[1].

The theory is based on the overall group:

$$G_{ST} = SU(3)_c \otimes SU(2)_L \otimes U(1)_Y. \quad (10.29)$$

G_{ST} corresponds to an *exact* local symmetry at the level of the Lagrangian, but spontaneously broken by the vacuum expectation value of the Higgs field, which reduces the symmetry according to:

$$SU(3)_c \otimes SU(2)_L \otimes U(1)_Y \to SU(3)_c \otimes U(1)_Q \quad (10.30)$$

where Q denotes the electric charge and $U(1)_Q$ the familiar gauge group of the electromagnetic interactions.

We can write the overall Lagrangian, $\mathcal{L}_{ST}$, as the sum over several terms:

$$\mathcal{L}_{ST} = \mathcal{L}_{ST,q} + \mathcal{L}_{ST,l} + \mathcal{L}_{ST,\phi} + \mathcal{L}_{ST,\phi-f} + \mathcal{L}_{ST,gauge} \quad (10.31)$$

where ϕ is the doublet of scalar fields in (6.5). We now consider explicitly the various terms.

Quarks.   In the notation of section 10.1, the colour and electroweak interactions of the quarks are described by the Lagrangian:

$$\mathcal{L}_{ST,q} = \sum_{f=1,3} \bar{Q}_f \, iD_\mu^{ST} \gamma^\mu \, Q_f + \sum_{f=1,3} \bar{U}_{Rf} \, iD_\mu^{ST} \gamma^\mu U_{Rf} +$$
$$+ \sum_{f=1,3} \bar{D}_{Rf} \, iD_\mu^{ST} \gamma^\mu D_{Rf} \quad (10.32)$$

---

[1] the name Standard Model was traditionally used, but we think it is time to recognize the theory status to the most sucessful extension of the Maxwell theory ever produced.

where now f represents the generation index ($f = 1$ for u, d, etc.); Q, U_R and D_R respectively the left-handed doublet and the right-handed singlet of each quark generation. The covariant derivatives are given by the expressions:

$$Q: iD_\mu^{ST} = i\partial_\mu - \left(g_S \sum_A \frac{\lambda^A}{2} g_\mu^A + g \sum_i \frac{\tau^i}{2} W_\mu^i + g' \frac{Y_Q}{2} B_\mu \right);$$

$$U_R, D_R: iD_\mu^{ST} = i\partial_\mu - \left(g_S \sum_A \frac{\lambda^A}{2} g_\mu^A + g' \frac{Y_{U,D}}{2} B_\mu \right);$$

$$Y_Q = \frac{1}{3}; \quad Y_U = +\frac{4}{3}; \quad Y_D = -\frac{2}{3}. \tag{10.33}$$

Leptons. Assuming the the neutrinos are purely left-handed, we can write:

$$\mathcal{L}_{ST,l} = \sum_{f=1,3} \bar{L}_f \, iD_\mu^{ST} \gamma^\mu \, L_f + \sum_{f=1,3} \bar{l}_{Rf} \, iD_\mu^{ST} \gamma^\mu l_{Rf} \tag{10.34}$$

where L_f represents the left-handed doublet of each generation ($f = 1$, $L = (\nu_e, e)_L$, etc.) and l_{fR} the right-handed singlet (e_R, etc.). The leptons are colour singlets, for which the covariant derivatives refer only to $SU(2)_L \otimes U(1)_Y$:

$$L: iD_\mu^{ST} = i\partial_\mu - \left(g \sum_i \frac{\tau^i}{2} W_\mu^i + g' \frac{Y_L}{2} B_\mu \right);$$

$$l_{fR}: iD_\mu^{ST} = i\partial_\mu - g' \frac{Y_l}{2} B_\mu;$$

$$Y_L = -1; \quad Y_l = -2. \tag{10.35}$$

Scalar doublet. The Lagrangian $\mathcal{L}_{ST,\phi}$ coincides with the Lagrangian $\mathcal{L}_{\phi W}$ constructed in chapter 6, equation (6.7). Explicitly:

$$\mathcal{L}_{ST,\phi} = (D_\mu \phi)^\dagger (D^\mu \phi) - V(\phi);$$
$$V(\phi) = \mu^2 \phi^\dagger \phi + \lambda (\phi^\dagger \phi)^2;$$
$$\lambda > 0, \quad \mu^2 < 0. \tag{10.36}$$

and the definition of the covariant derivative given in (6.5):

$$\phi: iD_\mu^{ST} = i\partial_\mu - \left(g \sum_i \frac{\tau^i}{2} W_\mu^i + g' \frac{Y_\phi}{2} B_\mu \right);$$

$$Y_\phi = +1. \tag{10.37}$$

Scalar doublet–fermion interactions. These interactions are given the name *Yukawa interactions*[2] and explicitly written:

$$\mathcal{L}_{ST,\phi-f} = \bar{Q}\, g^D \phi\, D_R + \epsilon_{ab} \bar{U}_R\, g^U\, Q^a \phi^b + \bar{L}\, g^l \phi\, l_R + h.c.$$

(10.38)

where g^D, g^U and g^l are (complex) matrices in the space of the flavour indices (which we have suppressed for notational simplicity) and $a, b = 1, 2$ are indices of weak isospin, $SU(2)_L$.

Gauge fields. Finally, cf. equation (6.2) and (8.22):

$$\mathcal{L}_{ST,gauge} = -\frac{1}{4}\sum_A G^A_{\mu\nu}(G^A)^{\mu\nu} - \frac{1}{4}\sum_i W^i_{\mu\nu}(W^i)^{\mu\nu} - \frac{1}{4}B_{\mu\nu}B^{\mu\nu};$$

$$SU(3)_c: \quad (G^A)^{\mu\nu} = \partial_\nu g^A_\mu - \partial_\mu g^A_\nu + g_S\, f^{ABC}\, g^B_\mu g^C_\nu;$$

$$SU(2): \quad W^i_{\mu\nu} = \partial_\nu W^i_\mu - \partial_\mu W^i_\nu + g\, \epsilon_{jik} W^j_\mu W^k_\nu;$$

$$U(1): \quad B_{\mu\nu} = \partial_\nu B_\mu - \partial_\mu B_\nu.$$

(10.39)

Note. The gauge interactions represented in (10.32) allow a global symmetry corresponding to mixing between the left-handed doublet Q_f, or the singlets $U_{R,f}$, or the singlets $D_{R,f}$. $SU(3)^Q_L \otimes SU(3)^U_R \otimes SU(3)^D_R$ is a very large group, a subgroup of the chiral flavour group which we introduced in chapter 8 as a global QCD symmetry, $SU(6)_L \otimes SU(6)_R$ in the case of six flavours.

An analogous symmetry for the lepton flavours is present in (10.34) and corresponds to the group $SU(3)^L_L \otimes SU(3)^l_R$.

These large global symmetries are explicitly broken by the Yukawa couplings of the scalar doublet to fermions, quarks and leptons, which reduce them to the only flavour symmetry of the Standard Theory which acts simultaneously on all the doublets and singlets, the symmetry $SU(2)_L \times U(1)_Y$, which is promoted to a gauge symmetry by the introduction of covariant derivatives. In its turn, $SU(2)_L \times U(1)_Y$ is reduced to the symmetry of electromagnetism, $U(1)_Q$, by the vacuum expectation value of the neutral component of the scalar doublet.

---

[2]The name is derived from the fact that the interaction has the form suggested originally by Yukawa for the pion-nucleon interaction.

CHAPTER 11

DIGRESSION: VIOLATION OF ISOTOPIC SPIN

CONTENTS

In this chapter we consider the currents associated with the chiral symmetry $SU(3)_L \otimes SU(3)_R$, generated by chiral transformations of the triplet of light quarks u, d and s.

In place of the chiral currents, we can use the vector and axial vector currents which, in the quark model, are written:

$$V^i = \bar{q}\frac{\lambda_i}{2}\gamma_\mu q; \tag{11.1}$$

$$A^i = \bar{q}\frac{\lambda_i}{2}\gamma_\mu\gamma_5 q. \tag{11.2}$$

In the limit of exact symmetry, the currents satisfy the continuity equations:

$$\partial^\mu J_\mu^A = 0, \ (J_\mu = V_\mu, A_\mu) \tag{11.3}$$

and the charges:

$$Q_A = \int J_0^A(\mathbf{x}, t)d^3\mathbf{x}^3;$$

$$[Q_A, Q_B] = if_{ABC}Q_C$$

are conserved.

We consider in addition the scalar and pseudoscalar densities (only light quarks):

$$S_i = \bar{q}\lambda_i q; \qquad P_i = i\bar{q}\lambda_i\gamma_5 q;$$

which satisfy the commutation rules:

$$[V^i, S_j] = if_{ijk}S_k; \quad [V^i, P_j] = if_{ijk}P_k;$$
$$[A^i, S_j] = id_{ijk}P_k; \quad [A^i, P_j] = id_{ijk}S_k;$$
$$[Q_A, \phi_i] = iT^A_{ij}\phi_j;$$
$$i, j = 0, 1, \cdots, 8. \tag{11.4}$$

11.1 GOLDSTONE BOSONS IN $SU(3)_L \otimes SU(3)_R$

We assume at the outset that the u, d and s quarks are massless.

If the strong interactions respect parity and strangeness conservation, the operators which can have a vacuum expectation value (VEV) are S_0, S_3, S_8. The symmetry is spontaneously broken into the subgroups:

$$SU(3)_L \otimes SU(3)_R \to_{S_0 \neq 0} SU(3)_V \to_{S_8 \neq 0} SU(2) \otimes U(1) \to$$
$$\to_{S_3 \neq 0} U(1)_Q \otimes U(1)_S. \tag{11.5}$$

The charges of the subgroup annihilate the vacuum. Let us examine the possible cases.

- $< 0|S_0|0 > \neq 0$: the relevant equations are:

$$\langle 0| [A^i, P_j] |0\rangle = id_{ijk} \langle 0|S_k|0\rangle =$$
$$= id_{ij0} \langle 0|S_0|0\rangle = i\sqrt{\frac{2}{3}}\delta_{ij} \langle 0|S_0|0\rangle. \tag{11.6}$$

All the axial charges are broken and we have an octet of *pseudoscalar* Goldstone bosons. The vector charges annihilate the vacuum; $SU(3)_V$ is an exact symmetry.

- As well as S_0, also $< 0|S_8|0 > \neq 0$: some vector charges do not any longer annihilate the vacuum and we have some scalar Goldstone bosons, with the quantum numbers of the K:

$$\langle 0| [V^i, S_j] |0\rangle = if_{ijk} \langle 0|S_k|0\rangle = if_{ij8} \langle 0|S_8|0\rangle. \tag{11.7}$$

- $< 0|S_3|0 > \neq 0$: other scalar Goldstone bosons appear with the quantum numbers of the pion.

In the real world, the lightest spin zero particle is the pion, a pseudoscalar; the real world seems to be close to the first case [36].

How close?

11.2   PIONS AND KAONS AS QUASI-GOLDSTONE BOSONS

In the world of strong interactions, there are no massless particles but the mass of the pion is very small on the scale of the nucleon masses, and the mass of the K meson is also rather small:

$$m_\pi^2 = 0.019 \text{ GeV}^2; \quad m_K^2 = 0.25 \text{ GeV}^2. \tag{11.8}$$

We make the following hypotheses:

- The vacuum breaks $SU(3)_L \otimes SU(3)_R$ to $SU(3)_V$: $<0|S_0|0> \neq 0$, with $<0|S_8|0>, <0|S_3|0> \approx 0$.

- Quarks have masses (m_u, m_d, m_s) small on the strong interaction scale.

The relations found in section 5.3 are modified because there are no longer poles at $p^2 = 0$ but they determine the masses of the pseudoscalar mesons in terms of the quark masses (which appear in the 4-divergence of the currents). The pseudoscalar mesons are *quasi-Goldstone bosons*, with small masses that vanish with the quark masses [37] [38].

We return to eq. (5.40) applied to the pseudoscalar density, P_i, but without assuming the vanishing of the divergence of the current:

$$q_\mu \int d^4x \; e^{iqx} < 0|T\left[J^\mu(x)P_i(0)\right]|0> =$$

$$= i \int d^4x \; e^{iqx} < 0|T\left[\partial_\mu J^\mu(x)P_i(0)\right]|0> +$$

$$+ i \int d^3x \; e^{-i\mathbf{qx}} < 0| \left[J^0(\mathbf{x},0)P_i(0)\right]|0> . \tag{11.9}$$

If we allow $q_\mu \to 0$ in (11.9), the left hand side tends to zero; there are no longer Goldstone bosons and therefore the correlation function no longer has a pole for $q^2 = 0$. Therefore we obtain the relation:

$$\int d^4x \; e^{iqx} < 0|T\left[\partial_\mu J^\mu(x)P_i(0)\right]|0> =$$

$$= - \int d^3x \; e^{-i\mathbf{qx}} < 0| \left[J^0(\mathbf{x},0)P_i(0)\right]|0> . \tag{11.10}$$

Now we focus on the case of the axial currents (11.2):

$$\int d^4x \; e^{iqx} < 0|T\left[\partial^\mu A_\mu^i(x)P_j(0)\right]|0> = -id_{ij0} < 0|S_0|0> . \tag{11.11}$$

We consider the possible cases, starting from $i = 1$. Explicitly[1]:

$$A_\mu^1 = \frac{1}{2}\left(\bar{u}\gamma_\mu\gamma_5 d + \bar{d}\gamma_\mu\gamma_5 u\right);$$

$$\partial^\mu A_\mu^1 = \frac{m_u + m_d}{2}P_1. \tag{11.12}$$

---

[1] In calculating the divergences of the current we use the Dirac equation $\partial^\mu \gamma_\mu q = -im_q q$, $(q = u, d)$ and its adjoint.

and therefore:

$$\frac{m_u + m_d}{2} \int d^4x \, e^{iqx} < 0|T\left(P^1(x)P^1(0)\right)|0> =$$

$$= -id_{110} < 0|S_0|0> = -iA \qquad (11.13)$$

where A is a constant. If we approximate the two-point function in (11.13) with the contribution of the one-particle state, the charged pion, we find:

$$\int d^4x \, e^{iqx} < 0|T\left(P^1(x)P^1(0)\right)|0> = Z_\pi^2 \frac{i}{q^2 - m_{\pi^+}^2} \qquad (11.14)$$

and equation (11.13) becomes (in the limit $q \to 0$):

$$\left(\frac{m_u + m_d}{2}\right) Z_\pi^2 \frac{-i}{q^2 - m_{\pi^+}^2} = -iA \text{ or}$$

$$m_u + m_d = Cm_{\pi^+}^2. \qquad (11.15)$$

Repeating for the cases (the particle which dominates the two-point function is indicated in parenthesis)

$$i = 3 \ (\pi^0), \quad i = 4 \ (K^+), \quad i = 6 \ (K^0), \qquad (11.16)$$

three similar equations are found:

$$m_u + m_d = Cm_{\pi^0}^2; \qquad (11.17)$$

$$m_u + m_s = Cm_{K^+}^2; \qquad (11.18)$$

$$m_d + m_s = Cm_{K^0}^2. \qquad (11.19)$$

We have assumed the the constants Z satisfy the symmetry conditions:

$$< 0|P^1|\pi^+> = < 0|P^3|\pi^0> = < 0|P^4|K^+> = < 0|P^6|K^0> \qquad (11.20)$$

which are appropriate in the zero mass limit of the light quarks, u, d, s in which the $SU(3)_V$ symmetry is exact.

We first consider the two equations which involve the sum $m_u + m_d$, that is the sum of (11.13) and (11.17) and the sum of (11.18) and (11.19). After having eliminated the unknown scale factor, C, we can determine the ratio:

$$\frac{m_u + m_d}{m_s + \frac{m_u + m_d}{2}} = \frac{m_{\pi^+}^2 + m_{\pi^0}^2}{m_{K^+}^2 + m_{K^0}^2} = R = 0.077. \qquad (11.21)$$

The smallness of the right-hand side indicates clearly that $(m_u + m_d)/2 \ll m_s$. To fix the mass of the s quark, we can go back to the mass spectrum of the baryon resonances considered in chapter 2; see Figure 2.4. Their quark composition and corresponding masses are summarised in Table 11.1. The linear behaviour of the masses with strangeness is interpreted naturally as

Table 11.1 Masses of the baryon resonances with different strangeness and maximum isotopic spin.

Name	Δ^{++}	Y^+	Ξ^0	Ω^-
quark content	(uuu)	(uus)	(uss)	(sss)
strangeness	0	-1	-2	-3
mass (MeV)	1232	1385	1530	1672
difference (MeV)	–	153	145	142

due to the m_s–m_u mass difference which, in view of the previous remark, is roughly equal to m_s.

If we take $m_s = 150$ MeV as an indicative value, we find:

$$\frac{m_u + m_d}{2} = 6 \text{ MeV}. \tag{11.22}$$

The result in (11.22) is remarkable. The mass scale of the *up* and *down* quarks fixes the scale of the breaking of $SU(2)_L \otimes SU(2)_R$ which is therefore broken at the MeV level, a surprisingly small value, given that the mass of the pion is about 140 MeV.

The explicit breaking represented by (11.22) implies that the mass of the pion should not be exactly zero, as required by the Goldstone theorem, but nevertheless very small compared to the strong interaction scale, for example the mass of the proton, which is closely associated with the value of $< 0|S_0|0 >$.

Put in another way, the parameter which characterises the breaking in vacuum of the chiral symmetry is of the order of the constant[2] $Z_\pi \simeq f_\pi \simeq 140$ MeV, while the explicit breaking of $SU(2)_L \otimes SU(2)_R$ is of order MeV, or the same scale as the breaking of isotopic spin symmetry, $SU(2)$.

11.3 VIOLATION OF ISOTOPIC SPIN SYMMETRY

To evaluate individually the masses of the u and d quarks, we must note that the π^+–π^0 and K^+–K^0 mass differences also contain electromagnetic corrections which correspond to the electrostatic energy of the charged particles[3]. Here we can use a result due to Dashen, according to which *the electromagnetic corrections cancel* in the particular combination of mass differences:

$$\left[m_{K^+}^2 - m_{K^0}^2\right] - \left[m_{\pi^+}^2 - m_{\pi^0}^2\right] = \Delta = -0.0052 \text{ GeV}^2.$$

---

[2] Not very different from the parameter which characterises QCD, Λ_{QCD}, equation (8.28).

[3] In particular, the mass difference of the quarks does not contribute to the π^+–π^0 mass difference which therefore is due entirely to electromagnetic effects.

We can therefore set this combination equal to what was obtained from the previous equations and put:

$$m_u - m_d = C\Delta. \tag{11.23}$$

The electromagnetic corrections, not calculable, affect other mass combinations, e.g. $m_{\pi^+}^2 - m_{\pi^0}^2$.

We therefore remain with (11.23) and with the two equations which involve the sum $m_u + m_d$, that is the sum of (11.13) and (11.17) and the sum of (11.18) and (11.19). From here, we obtain the ratio (11.21) and:

$$\frac{m_u - m_d}{m_s + \frac{m_u + m_d}{2}} = 2\frac{\left(m_{K^+}^2 - m_{K^0}^2\right) - \left(m_{\pi^+}^2 - m_{\pi^0}^2\right)}{\left(m_{K^+}^2 + m_{K^0}^2\right)} = -0.021 \tag{11.24}$$

from which, finally:

$$\left[\begin{array}{c} \frac{m_u}{m_s} = 0.029 \\ \frac{m_d}{m_s} = 0.051 \end{array} \right]. \tag{11.25}$$

If we again take $m_s = 150$ MeV, we find:

$$m_u = 4.4 \text{ MeV}, \tag{11.26}$$

$$m_d = 7.7 \text{ MeV}. \tag{11.27}$$

Another surprising result. The mass difference of the *up* and *down* quarks, which breaks isotopic spin symmetry, is actually not small compared to the masses themselves [39]. How is it that isotopic spin is a symmetry so well respected by the strong interactions? The explanation lies in the smallness of the masses of the light quarks; the real world is close to isotopic spin symmetry not because the masses are very similar, like the masses of the neutron and proton, but because both are roughly equal to zero!

A final comment on isotopic spin. The initial expectation, hinted at in chapter 2, was that the symmetry was broken only by electromagnetic interactions, which obviously distinguish the proton from the neutron (and the *up* quark from the *down* quark). We now see that to this effect should be added another associated with the quark masses and not controlled by purely electromagnetic interactions.

The new effect is necessary to explain the phenomenology of isotopic spin violation. Actually, all attempts to calculate the proton–neutron mass difference based purely on electromagnetism lead (as is natural) to a proton heavier than the neutron, with an M_p–M_n difference equal to about +1.5 MeV. Since $p = (uud)$ and $n = (udd)$, the new effect adds m_d–$m_u \simeq 3$ MeV to the mass of the neutron, bringing it into agreement with experimental data.

More precise analyses, carried out at the end of the 1960s, confirm the agreement of the phenomenology with the presence of *two components* in isotopic spin violation: electromagnetic effects and the m_u–m_d mass difference, which are (by chance, as far as we know) of the same order of magnitude.

CHAPTER 12

MIXING and C P VIOLATION IN NEUTRAL K, B and D MESONS

CONTENTS

12.1   THE K^0–$\bar{K}^0$ SYSTEM

In this section we illustrate the characteristic phenomenology of the violation of CP symmetry in the decays of neutral K mesons.

Hamiltonian for the neutral K mesons.   In the rest system of the meson, the Hamiltonian reduces to a 2×2 matrix which we write in the basis:

$$K^0 = \begin{pmatrix} 1 \\ 0 \end{pmatrix} ; \quad \bar{K}^0 = \begin{pmatrix} 0 \\ 1 \end{pmatrix} ; \tag{12.1}$$

$$\mathbf{M} = \begin{pmatrix} M_{11} & M_{12} \\ M_{21} & M_{22} \end{pmatrix} . \tag{12.2}$$

To complete the description of neutral K mesons, we must take account of the fact that these are unstable particles. Similarly to the case of the Z^0, Section 7.4, we write the total Hamiltonian in the rest system:

$$\mathbf{H} = \mathbf{M} - \frac{i}{2}\mathbf{\Gamma} \tag{12.3}$$

where Γ is the Hermitian matrix which describes the overall set of K^0 and $\bar{K}^0$ decays:

$$\mathbf{\Gamma} = \begin{pmatrix} \Gamma_{11} & \Gamma_{12} \\ \Gamma_{21} & \Gamma_{22} \end{pmatrix};$$

$$\Gamma_{11}, \Gamma_{22} \text{ real}, \quad \Gamma_{21} = \Gamma_{12}^*. \tag{12.4}$$

Finally, we write the matrix $\mathbf{H}$ in the form:

$$\mathbf{H} = \begin{pmatrix} h & l \\ m & n \end{pmatrix} \tag{12.5}$$

with complex $h, \cdots n$, and we determine the eigenvalues and eigenvectors to connect to the quantities physically measurable in the decays of neutral K mesons (for an analysis of CP and CPT symmetries in these decays, see [40], [35] and [41]).

CP, T and CPT symmetries. We choose the phases of the kets $|K^0 >$ and $|\bar{K}^0 >$ so that:

$$CP|K^0 >= -|\bar{K}^0 >, \quad CP|\bar{K}^0 >= -|K^0 > . \tag{12.6}$$

(The convention used here is opposite to the one used in [40] for reasons which will be explained in Chapter 13.)

In the basis (12.1), the CP transformation acts on the Hermitian matrices $\mathbf{M}$ and $\mathbf{\Gamma}$ through the Pauli matrix $-\tau_1$:

$$CP : \mathbf{X} \to \tau_1 \mathbf{X} \tau_1 \quad (\mathbf{X} = \mathbf{M}, \mathbf{\Gamma}) \tag{12.7}$$

while time inversion, T, corresponds to taking the complex conjugate:

$$T : \mathbf{X} \to \mathbf{X}^*, \tag{12.8}$$

from which:

$$CPT : \mathbf{X} \to \tau_1 \mathbf{X}^* \tau_1. \tag{12.9}$$

The symmetry under CP is respected by the terms in $\mathbf{M}$ or $\mathbf{\Gamma}$ proportional to the Pauli matrices $\mathbf{1}$ and τ_1, and is violated by terms proportional to τ_2, like the imaginary part of δM^2 in (12.2), and to τ_3, while CPT symmetry requires the matrices $\mathbf{1}$, τ_1 (real and commuting with τ_1) and τ_2 (imaginary and anticommuting with τ_1), but not τ_3 (real and anticommuting with τ_1; cf. Table 12.1).

Invariance under TCP requires equality of the diagonal terms in (12.2) and (12.4). CPT invariance is satisfied as a consequence of the TCP theorem, in the description of the K mesons with local fields; cf. Section 13.2. We can however keep two different masses and widths on the diagonal in $\mathbf{H}$ and maintain the CP and CPT symmetries separately to compare with the data.

Table 12.1 Symmetry properties of the terms in the expansion $\mathbf{X} = c_0\mathbf{1} + c_1\tau_1 + c_2\tau_2 + c_3\tau_3$ ($\mathbf{X} = \mathbf{M}, \mathbf{\Gamma}$) under CP, T and CPT transformations. The sign $+$ corresponds to invariance under the corresponding symmetry.

	c_0	c_1	c_2	c_3
CP	$+$	$+$	$-$	$-$
T	$+$	$+$	$-$	$+$
CPT	$+$	$+$	$+$	$-$

Eigenvalues. They are the solutions to the characteristic equation:

$$Det \begin{pmatrix} h - \lambda & l \\ m & n - \lambda \end{pmatrix} \tag{12.10}$$

from which follow the two solutions:

$$\lambda_\pm = \frac{h + n \pm \sqrt{(h - n)^2 + 4lm}}{2}. \tag{12.11}$$

Eigenvectors. We write the eigenvectors corresponding to $\lambda_\pm$ in the form:

$$v_\pm = \begin{pmatrix} p_\pm \\ q_\pm \end{pmatrix} \tag{12.12}$$

and we easily find:

$$\left(\frac{q}{p}\right)_\pm = \frac{n - h \pm \sqrt{(h - n)^2 + 4lm}}{2l}. \tag{12.13}$$

In the limit in which the CP and CPT symmetries are exact, the matrices $\mathbf{M}$ and $\mathbf{\Gamma}$ are combinations of $\mathbf{1}$ and τ_1, which correspond to $h = n$ and $l = m$. In this case, the eigenvectors are the simple combinations:

$$CP, \ CPT : \text{ exact symmetries;}$$

$$v_+ = \frac{1}{\sqrt{2}} \begin{pmatrix} 1 \\ 1 \end{pmatrix} = \frac{K^0 + \bar{K}^0}{\sqrt{2}} = K_2 \quad (CP = -1);$$

$$v_- = \frac{1}{\sqrt{2}} \begin{pmatrix} 1 \\ -1 \end{pmatrix} = \frac{K^0 - \bar{K}^0}{\sqrt{2}} = K_1 \quad (CP = +1). \tag{12.14}$$

The K_1 state can decay into two π mesons (which in the state with $L = 0$ have $CP = +1$) while the K_2 state must decay into at least three π mesons

($CP = -1$ in $L = 0$) with a much reduced phase space. This situation persists even in the presence of a small CP violation and the notation K_S and K_L (S = short, L = long) is used for the two eigenstates of $\mathbf{M}$ which are close to K_1 and K_2. Experimentally $\tau(K_L)/\tau(K_S) \sim 500$. To demonstrate this closeness, this can be written

$$CP, \ CPT : \text{ general violations};$$ (12.15)

$$v_+ = K_L = N_L(K_2 + \epsilon_L K_1);$$
$$v_- = K_S = N_S(K_1 + \epsilon_S K_2)$$ (12.16)

where $N_{S,L}$ are normalisation factors. Comparing (12.12) and (12.13), we find:

$$(\frac{q}{p})_+ = \frac{n - h + \sqrt{(h-n)^2 + 4lm}}{2l} = \frac{1 - \epsilon_L}{1 + \epsilon_L};$$
$$(\frac{q}{p})_- = \frac{n - h - \sqrt{(h-n)^2 + 4lm}}{2l} = -\frac{1 - \epsilon_S}{1 + \epsilon_S}.$$ (12.17)

Notation often used in the literature is:

$$\epsilon = \frac{\epsilon_S + \epsilon_L}{2}; \qquad \Delta = \frac{\epsilon_S - \epsilon_L}{2}.$$ (12.18)

The measurable quantities connected to the matrix $\mathbf{H}$ are the masses and the widths of the $K_{S,L}$ states and the parameters ϵ and Δ; cf. Tables 12.2 and 12.3, and [5].

Table 12.2   Masses (MeV) and lifetimes (s) of neutral K mesons.

$(m_S + m_L)/2$	$\Delta m = m_L - m_S$	τ_S	τ_L
497.614 ± 0.024	$(3.483 \pm 0.006) \cdot 10^{-12}$	(0.8937 ± 0.0012) $\times 10^{-10}$	(511.6 ± 2.1) $\times 10^{-10}$

Table 12.3 CP and CPT violating parameters in the mass matrix of the neutral K mesons.

$|\epsilon|$	$\arg \epsilon$	$\text{Re}(\Delta)$	$\text{Im}(\Delta)$
$(2.228 \pm 0.011) \cdot 10^{-3}$	$(43.51 \pm 0.05)°$	$(25 \pm 23) \cdot 10^{-5}$	$(-0.6 \pm 1.9) \cdot 10^{-5}$

The quantity Δm can also be given in units of (time)$^{-1}$:

$$\Delta m = m_L - m_S = (0.5290 \pm 0.0015) \times 10^{10} \text{ s}^{-1}. \qquad (12.19)$$

To obtain the parameters which appear in the Hamiltonian, **H**, from measurable quantities it is necessary to invert equations (12.11) and (12.17). There are four complex relations which determine the four complex elements of **H**. The inversion of these equations is easily done if we limit ourselves to terms of first order in the (in any case small) violations of CP and CPT.

In this approximation, the square root in (12.11) simplifies to:

$$\left[(h-n)^2 + 4lm\right]^{\frac{1}{2}} \simeq 2\left[(M_{12} - \frac{i}{2}\Gamma_{12})(M_{12}^* - \frac{i}{2}\Gamma_{12}^*)\right]^{\frac{1}{2}} =$$

$$= 2\left[|M_{12}|^2 - \frac{1}{4}|\Gamma_{12}|^2 - iRe(M_{12}\Gamma_{12}^*)\right]^{\frac{1}{2}} \sim 2(ReM_{12} - \frac{i}{2}Re\Gamma_{12}) \quad (12.20)$$

and we find:

$$\lambda_\pm = \frac{M_{11} + M_{22} - \frac{i}{2}(\Gamma_{11} + \Gamma_{22}) \pm 2(ReM_{12} - \frac{i}{2}Re\Gamma_{12})}{2} \qquad (12.21)$$

or:

$$M_{11} + M_{22} = m_S + m_L;$$
$$\Gamma_{11} + \Gamma_{22} = \Gamma_S + \Gamma_L; \qquad (12.22)$$
$$2Re(M_{12}) = m_L - m_S = +\Delta m;$$
$$2Re(\Gamma_{12}) = -(\Gamma_S - \Gamma_L) = -\Delta\Gamma. \qquad (12.23)$$

We now consider equation (12.17). Taking the product on both sides of the two equations, we find:

$$-\frac{m}{l} = -(\frac{1-\epsilon_S}{1+\epsilon_S})(\frac{1-\epsilon_L}{1+\epsilon_L}) \sim -(1 - 2(\epsilon_S + \epsilon_L)) = -(1 - 4\epsilon) \qquad (12.24)$$

while expanding the ratio on both sides gives:

$$\frac{\sqrt{(h-n)^2 + 4lm} + n - h}{\sqrt{(h-n)^2 + 4lm} - (n-h)} \sim 1 - \frac{h-n}{\sqrt{lm}} =$$

$$= (\frac{1+\epsilon_S}{1-\epsilon_S})(\frac{1-\epsilon_L}{1+\epsilon_L}) \sim 1 + 2(\epsilon_S - \epsilon_L) = 1 + 4\Delta. \qquad (12.25)$$

For the left hand side of (12.24) we find:

$$\frac{m}{l} = \frac{M_{12}^* - \frac{i}{2}\Gamma_{12}^*}{M_{12} - \frac{i}{2}\Gamma_{12}} = \frac{1 + (-iImM_{12} - \frac{1}{2}Im\Gamma_{12})(ReM_{12} - \frac{i}{2}Re\Gamma_{12})^{-1}}{1 - (-iImM_{12} - \frac{1}{2}Im\Gamma_{12})(ReM_{12} - \frac{i}{2}Re\Gamma_{12})^{-1}} =$$

$$= 1 + 2\frac{(-iImM_{12} - \frac{1}{2}Im\Gamma_{12})}{(ReM_{12} - \frac{i}{2}Re\Gamma_{12})} = 1 + 2\frac{(ImM_{12} - \frac{i}{2}Im\Gamma_{12})}{(iReM_{12} + \frac{1}{2}Re\Gamma_{12})}. \qquad (12.26)$$

Using (12.23) we can express the real parts in the denominator in terms of Δm and $\Delta\Gamma$:

$$\frac{m}{l} = 1 + \frac{4}{\Delta m}\frac{(ImM_{12} - \frac{i}{2}Im\Gamma_{12})}{(-i\frac{2\Delta m}{\Delta\Gamma} + 1)}(\frac{2\Delta m}{\Delta\Gamma}). \tag{12.27}$$

The term $Im\Gamma_{12}$ depends on the direct CP violation in weak decays, which we know experimentally to be around 10^{-3} times the violation in mixing represented by ϵ. With $Im\Gamma_{12} \sim 0$, we find:

$$\epsilon = \frac{1}{\Delta m}\frac{ImM_{12}}{(-i\tan\phi_{SW} + 1)}(\tan\phi_{SW}) = \frac{ImM_{12}}{\Delta m}(\sin\phi_{SW})e^{+i\phi_{SW}} \tag{12.28}$$

where we have introduced the so-called *superweak phase*[1], ϕ_{SW}, defined by:

$$\phi_{SW} = \arctan[\frac{2\Delta m}{\Delta\Gamma}] = 43.45°. \tag{12.29}$$

From (12.28) we see that the argument of the complex number ϵ is fixed by the value of ϕ_{SW}, independently of the details of the electroweak theory which are contained in the quantity ImM_{12}, and it is in excellent agreement with the experimental value; cf. Table 12.3.

To finish, we analyse equation (12.25), which leads directly to the relation:

$$\Delta = -\frac{1}{4}\frac{M_{11} - M_{22} - \frac{i}{2}(\Gamma_{11} - \Gamma_{22})}{ReM_{12} - \frac{i}{2}Re\Gamma_{12}} =$$
$$= \frac{i}{\Delta\Gamma}\frac{M_{11} - M_{22} - \frac{i}{2}(\Gamma_{11} - \Gamma_{22})}{-i\frac{2\Delta m}{\Delta\Gamma} + 1} =$$
$$= \frac{i}{2}(\sin\phi_{SW})e^{+i\phi_{SW}}\frac{M_{11} - M_{22} - \frac{i}{2}(\Gamma_{11} - \Gamma_{22})}{\Delta m} \tag{12.30}$$

which connects the CPT violation represented by $M_{11} - M_{22}$ and $\Gamma_{11} - \Gamma_{22}$ to the observable quantity Δ. In particular, for the $K^0 - \bar{K}^0$ mass difference, we find:

$$|\frac{M_{K^0} - M_{\bar{K}^0}}{M_K}| = \frac{4\Delta m}{\sin(2\phi_{SW})M_K}|Re\Delta| = (7.0 \pm 7.0) \times 10^{-18}$$

$$(TCP, \text{ strange particles}), \tag{12.31}$$

an extraordinary confirmation of the CPT theorem (see [1] and [41]) which exceeds even the already very precise limit for stable matter obtained with

---

[1]This would be the argument of ϵ in a theory in which the electroweak interactions exactly respected CP and the observed violation were to first order the effect of a *superweak interaction* [44], which produced the K^0–$\bar{K}^0$ transition as its only observable effect. From this comes the name attributed to the phase (12.29).

the experiments ATRAP [42] and ASACUSA [43] at CERN[2]:

$$\left|\frac{M_p - M_{\bar{p}}}{M_p}\right| < 0.9 \cdot 10^{-10} \qquad (TCP, \text{ stable matter}). \qquad (12.32)$$

To conclude this section, we write down the relationships of M_{12} with the observables of the K meson mass matrix:

$$Re(M_{12}) = \frac{\Delta m}{2}; \qquad (12.33)$$

$$\frac{Im(M_{12})}{\Delta m} = \frac{Re(\epsilon)}{\sin \phi_{SW} \cos \phi_{SW}} = \frac{|\epsilon|}{\sin \phi_{SW}}. \qquad (12.34)$$

The CKM predictions for these parameters will be shown in chapter 13.

12.2 MIXING OF NEUTRAL B AND D MESONS

In general, the formalism which we have shown for K mesons remains valid for the case of mixing between $B^0 = (\bar{b}d)$ and $\bar{B}^0 = (b\bar{d})$ mesons. However, an important difference occurs at the matrix level for the widths. The weak decays of B mesons are dominated by the decay of the heavy quark, the b quark, and there are no significant differences in decay rates into different CP channels, as is the case for K mesons. A reasonable approximation for B mesons is to neglect completely the off-diagonal width terms:

$$M_{12}(B) >> \Gamma_{12}(B) \simeq 0. \qquad (12.35)$$

In this case, even if M_{12} is a complex number, in the convention used for the CKM matrix, the phase can always be eliminated with a suitable choice of phases for the B^0 and $\bar{B}^0$ states. This means that a definition of CP symmetry which is conserved in the mass matrix exists and the only possible observable is the mass difference between the two eigenstates, B_H and B_L ($H = $ heavy, $L = $ light), which is given by:

$$|\Delta M(B)| = 2|M_{12}|. \qquad (12.36)$$

The same considerations hold for the $D^0 = (c\bar{u})$ and $\bar{D}^0 = (u\bar{c})$ mesons.

---

[2]The result obtained by the ATRAP experiment at CERN is based on measurements of q/m for protons and antiprotons, interacting with the same electromagnetic fields; a similar precision was obtained by the ASAKUSA experiment, by measuring the ratio of the mass of the antiproton to that of the electron.

FLAVOUR–CHANGING NEUTRAL CURRENTS

CONTENTS

The expression *flavour–changing neutral current* processes (FCNC) denotes semileptonic processes with a change of flavour but without net charge transfer to the leptons, or non-leptonic processes which should occur at higher order in the electroweak theory, as a consequence of the fact that the neutral current does not produce a change of flavour to lowest order; cf. equation (10.16). Well known examples of FCNC processes are the decay $K_L \to \mu^+\mu^-$ and the $\Delta S = 2$ transition $\bar{K}^0$–K^0. In the electroweak theory, the FCNC amplitudes due to exchange of the intermediate bosons, W and Z, should be finite, as a consequence of the fact that it is not possible to add flavour-changing counterterms to the neutral current.

The convergence of the integration over internal momenta is ensured by the GIM mechanism, or by its extension to the six quark theory. This has a consequence that the relevant internal momenta are of order or larger than the mass of the charm quark. The process is therefore dominated by *short distance* amplitudes and it is possible to calculate the weak contribution without taking account of strong interaction corrections, as a result of asymptotic freedom of the colour interactions.

If there are no further contributions dominated by long distances, and thus affected by effectively non-calculable corrections due to strong interactions, eventual discrepancies of experimental data with the amplitudes calculated in the standard theory can give significant indications of new physics beyond the Standard Theory. The FCNC processes dominated by short distances have in recent years become an effective instrument to test the Standard Theory at energies not yet accessible to available accelerators (for a recent discussion see [45]).

Despite its historic role as an FCNC prototype, the $K_L \to \mu^+\mu^-$ decay does actually not lend itself to provide precision information since the process can occur also via the $K_L \to \gamma\gamma \to \mu^+\mu^-$ channel, which is dominated by long distance contributions.

Experimentally:

$$B(K_L \to \gamma\gamma) \simeq 6 \cdot 10^{-4}; \tag{13.1}$$

$$B(K_L \to \mu\mu) \simeq 7 \cdot 10^{-9}. \tag{13.2}$$

Since the probability of $\gamma\gamma \to \mu^+\mu^-$ is of order $(\alpha/\pi)^2 \simeq 4 \cdot 10^{-6}$, it can be seen that the process through the intermediate 2γ state can account for a significant part of the process amplitude.

In this chapter, we calculate in an illustrative way the weak amplitudes for transitions with a double change of flavour, $\Delta F = 2$, processes in which the weak amplitude is dominant and for which searches in different laboratories are under way.

13.1   THE $\bar{K}^0 \to K^0$ AMPLITUDE

At quark level, the $\bar{K}^0 \to K^0$ transition is described by the process:

$$s + \bar{d} \to d + \bar{s} \tag{13.3}$$

and by Feynman diagrams shown in Figure 13.1.

If we take the external momenta to be zero, only one momentum runs in the loop, which we denote as q. The internal fermion line is a superposition of propagators of the u, c and t quarks.

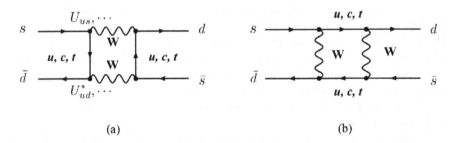

(a)   (b)

Figure 13.1   Feynman diagrams for the $\bar{K}^0 \to K^0$ transition.

Including also the neighbouring vertices, we have, for the left line in the diagram of Figure 13.1(a), a term[1] $\bar{v}_d S(q)_{\mu\nu} u_s$ with:

$$S(q)_{\mu\nu} = \gamma_\mu(1-\gamma_5)\left[a\frac{1}{\not{q}-m_u} + b\frac{1}{\not{q}-m_c} + c\frac{1}{\not{q}-m_t}\right]\gamma_\nu(1-\gamma_5);$$

$$a = U_{ud}^* U_{us}, \text{ etc.}$$

$$a + b + c = 0. \tag{13.4}$$

The relation $a = -b - c$ is equivalent to subtracting the u quark propagator from the propagators of the c and t, which makes them more convergent. With obvious manipulations, having set $m_u^2 = 0$, we find:

$$S(q)_{\mu\nu} = 2\gamma_\mu\not{q}\left[b\frac{m_c^2}{q^2(q^2-m_c^2)} + c\frac{m_t^2}{q^2(q^2-m_t^2)}\right]\gamma_\nu(1-\gamma_5). \tag{13.5}$$

We can therefore write the amplitude corresponding to the diagram of Figure 13.1(a) as follows[2]:

$$A^{(a)} = \int\frac{d^4q}{(2\pi)^4} \times 4\left[\frac{g^2}{8}\right]^2 [\bar{v}_d\gamma_\mu\not{q}\gamma_\nu(1-\gamma_5)u_s][\bar{u}_d\gamma_\sigma\not{q}\gamma_\rho(1-\gamma_5)v_s] \times$$

$$\times\left[b^2\frac{m_c^4}{q^4(q^2-m_c^2)^2} + 2bc\frac{m_c^2 m_t^2}{q^4(q^2-m_c^2)(q^2-m_t^2)} + c^2\frac{m_t^4}{q^4(q^2-m_t^2)^2}\right] \times$$

$$\times\left(-g^{\mu\rho} + \frac{q^\mu q^\rho}{M_W^2}\right)\left(-g^{\nu\sigma} + \frac{q^\nu q^\sigma}{M_W^2}\right)\frac{1}{(q^2-M_W^2)^2}. \tag{13.6}$$

The product of the numerators in the W propagators gives rise to four terms, according to whether we take the first or second term in the first or second numerator. We also keep in mind that the symmetric integration in q means that:

$$\int d^4q \, q^\alpha q^\beta \, F(q^2) = \frac{1}{4}g^{\alpha\beta}\int d^4q \, q^2 \, F(q^2). \tag{13.7}$$

The term which corresponds to $g^{\mu\rho}g^{\nu\sigma}$ gives rise to a product of three γ matrices, which can be simplified using the formula given in [1]:

$$\gamma^\nu\gamma^\alpha\gamma^\mu = g^{\nu\alpha}\gamma^\mu + g^{\alpha\mu}\gamma^\nu - g^{\nu\mu}\gamma^\alpha - i\epsilon^{\nu\alpha\mu\rho}\gamma_\rho\gamma_5 \tag{13.8}$$

---

[1] We recall the Feynman rules for the spinors: $u(p)$ $(\bar{v}(p))$ for a particle (antiparticle) in the initial state and $\bar{u}(p)$ $(v(p))$ for a particle (antiparticle) in the final state; the suffices d and s specify the flavour; the different terms are written tracing the fermion line following the arrows.

[2] The factor i for each vertex and for every propagator gives a factor $i^8 = 1$. The factor $g^2/8$ associated with every W line is reconstructed from formula (4.17), which gives the interaction with the doublets in the following way. A factor $1/4$ comes from the fact that the left-handed fields appear in (4.17); therefore in each vertex there is a factor $1/2(1-\gamma_5)$. In addition, the $W^{1,2}$ fields are coupled with the Pauli matrices $\tau^{1,2}/2$, while the vertices in Figure 13.1 correspond to the raising and lowering operators, $\tau^\pm$. The relation $\tau^1/2 \otimes \tau^1/2 + \tau^2/2 \otimes \tau^2/2 = 1/2\tau^+ \otimes \tau^- + \cdots$ provides the final factor $\frac{1}{2}$.

and the one obtained by lowering the indices:

$$\gamma_\mu \gamma_\alpha \gamma_\nu = g_{\nu\alpha}\gamma_\mu + g_{\alpha\mu}\gamma_\nu - g_{\nu\mu}\gamma_\alpha - i\epsilon_{\mu\alpha\nu\rho}\gamma^\rho\gamma_5. \tag{13.9}$$

Substituting equation (13.8) we find:

$$\gamma_\mu \gamma_\alpha \gamma_\nu (1 - \gamma_5) \otimes \gamma^\nu \gamma^\alpha \gamma^\mu (1 - \gamma_5) =$$
$$= 10\gamma_\sigma (1 - \gamma_5) \otimes \gamma^\sigma (1 - \gamma_5) - i\epsilon^{\mu\alpha\nu\rho}\gamma_\mu\gamma_\alpha\gamma_\nu (1 - \gamma_5) \otimes \gamma_\rho (1 - \gamma_5). \tag{13.10}$$

Now we use equation (13.9) and find:

$$-i\epsilon^{\mu\alpha\nu\rho}\gamma_\mu\gamma_\alpha\gamma_\nu (1 - \gamma_5) \otimes \gamma_\rho (1 - \gamma_5) =$$
$$= -6\gamma^\sigma (1 - \gamma_5) \otimes \gamma_\sigma (1 - \gamma_5) \tag{13.11}$$

or:

$$\gamma_\mu \gamma_\alpha \gamma_\nu (1 - \gamma_5) \otimes \gamma^\nu \gamma^\alpha \gamma^\mu (1 - \gamma_5) = 4\gamma_\mu (1 - \gamma_5) \otimes \gamma^\mu (1 - \gamma_5). \tag{13.12}$$

In the other terms, the relations $\not{q}\not{q} = q^2$ and $\not{q}\not{q}\not{q} = q^2\not{q}$ straightaway simplify the product of three γ matrices.

$$A^{(a)} = \int \frac{d^4q}{(2\pi)^4} \times 4 \left[\frac{g^2}{8}\right]^2 [\bar{v}_s\gamma_\mu (1 - \gamma_5)u_d] [\bar{u}_s\gamma^\mu (1 - \gamma_5)v_d] \times$$
$$\times \left[b^2 \frac{m_c^4}{q^2(q^2 - m_c^2)^2} + 2bc\frac{m_c^2 m_t^2}{q^2(q^2 - m_c^2)(q^2 - m_t^2)} + c^2 \frac{m_t^4}{q^2(q^2 - m_t^4)}\right] \times$$
$$\times \left(1 - 2\frac{q^2}{M_W^2} + \frac{1}{4}\frac{q^4}{M_W^4}\right)\right) \frac{1}{(q^2 - M_W^2)^2}. \tag{13.13}$$

Naturally we must remember that in the denominator of the propagators the Feynman rule holds which implies the substitution $m^2 \to m^2 - i\epsilon$ for every mass m^2. In this way we can turn the integration in q^0 from $-\infty$ to $+\infty$ into an integration along the imaginary axis, setting $q^0 = iq^4$ and integrating q^4 between $-\infty$ and $+\infty$ (Wick rotation). Under this transformation:

$$q^2 \to -q^2 \tag{13.14}$$

and furthermore, in spherical coordinates:

$$d^4q \to i\pi^2 q^2 dq^2. \tag{13.15}$$

Putting everything together and recalling that $\frac{g^2}{8M_W^2} = \frac{G_F}{\sqrt{2}}$, we can write:

$$A^{(a)} = -i\frac{G_F^2 M_W^2}{8\pi^2} \times \sum_{i,j=c,t} C_i C_j E(x_i, x_j) \times$$
$$\times [\bar{v}_s\gamma_\mu (1 - \gamma_5)u_d] [\bar{u}_s\gamma^\mu (1 - \gamma_5)v_d] \tag{13.16}$$

where $x_i = m_i^2/M_W^2$, $C_i = U_{id}^* U_{is}$, and:

$$E(x_i, x_j) = x_i x_j \int_0^\infty dt \left(1 + 2t + \frac{1}{4}t^2\right) \frac{1}{(t + x_i)(t + x_j)} \frac{1}{(t + 1)^2}. \quad (13.17)$$

Reasoning similarly, from the diagram of Figure 13.1 (b) we find:

$$A^{(b)} = -i\frac{G_F^2 M_W^2}{8\pi^2} \times \sum_{i,j=c,t} C_i C_j E(x_i, x_j) \times$$

$$\times \left[\bar{v}_s \gamma_\mu (1 - \gamma_5) v_d\right] \left[\bar{u}_s \gamma^\mu (1 - \gamma_5) u_d\right]. \quad (13.18)$$

The functions $E(x_i, x_j)$ have been calculated by Inami and Lim [46]. For us, a few specific examples which we can recalculate directly will suffice.

Calculation of the integrals. The integrals (13.17) reduce to two integrals:

$$\mathcal{I}(a, b) = \int_0^\infty dt \, \frac{1}{(t + a)(t + b)}; \quad (13.19)$$

$$\mathcal{I}(a, b, c) = \int_0^\infty dt \, \frac{1}{(t + a)(t + b)(t + c)}. \quad (13.20)$$

The first is immediately calculated:

$$\mathcal{I}(a, b) = \frac{1}{b - a} \ln \frac{b}{a}; \quad (13.21)$$

$$\mathcal{I}(a, a) = \frac{1}{a}. \quad (13.22)$$

We also note:

$$\mathcal{I}_1 = \int_0^\infty dt \, \frac{1}{(t + a)^2 (t + b)} = -\frac{\partial}{\partial a} \mathcal{I}(a, b) =$$

$$= \frac{1}{(b - a)} \left[-\frac{1}{(b - a)} \ln \left(\frac{b}{a}\right) + \frac{1}{a}\right];$$

$$\mathcal{I}_2 = \int_0^\infty dt \, \frac{1}{(t + a)^2 (t + b)^2} = \frac{\partial}{\partial b} \frac{\partial}{\partial a} \mathcal{I}(a, b) =$$

$$= \frac{1}{(b - a)^2} \left[-\frac{2}{(b - a)} \ln \left(\frac{b}{a}\right) + \frac{1}{b} + \frac{1}{a}\right]. \quad (13.23)$$

In the second case, the fraction is decomposed according to the formula:

$$\frac{1}{(t + a)(t + b)(t + c)} = \frac{\alpha}{(t + a)} + \frac{\beta}{(t + b)} + \frac{\gamma}{(t + c)},$$

$$\alpha = \frac{1}{(b - a)(c - a)}; \quad \beta = \frac{1}{(a - b)(c - b)}; \quad \gamma = \frac{1}{(a - c)(b - c)};$$

$$\alpha + \beta + \gamma = 0. \quad (13.24)$$

from which, integrating between 0 and $\lambda \gg a, b, c$

$$\mathcal{I}(a, b, c) = \alpha \ln \frac{\lambda}{a} + \beta \ln \frac{\lambda}{b} + \gamma \ln \frac{\lambda}{c} =$$

$$= \frac{1}{(b - a)(c - a)} \ln \left(\frac{c}{a}\right) + \frac{1}{(a - b)(c - b)} \ln \left(\frac{c}{b}\right). \quad (13.25)$$

We note:

$$\mathcal{I}_2(a, b, c) = \int_0^\infty dt \, \frac{1}{(t + a)(t + b)(t + c)^2} = -\frac{\partial}{\partial c} \mathcal{I}(a, b, c) =$$

$$= \frac{1}{(b - a)} \left[\frac{1}{(c - a)^2} \ln \left(\frac{c}{a}\right) - \frac{1}{(c - b)^2} \ln \left(\frac{c}{b}\right)\right] + \frac{1}{c(c - a)(c - b)}. \quad (13.26)$$

Results. From the preceding formulae, we find:

$$E(x_t, x_t) = x_t^2 \int_0^\infty dt \left(1 + 2t + \frac{1}{4}t^2\right) \frac{1}{(t + x_t)^2} \frac{1}{(t + 1)^2} =$$

$$= x_t^2 \int_0^\infty dt \left(\frac{1}{4}\mathcal{I}(x_t, 1) + \frac{3}{2}\mathcal{I}_1(x_t, 1) - \frac{3}{4}\mathcal{I}_2(x_t, 1)\right). \quad (13.27)$$

From here, we obtain:

$$E(x_t, x_t) = x_t^2 \left(\frac{1}{4}\mathcal{I}_0 + \frac{3}{2}\mathcal{I}_1 - \frac{3}{4}\mathcal{I}_2\right)$$

$$= \frac{x_t}{4} + \frac{3}{2}\frac{x_t^2}{(x_t - 1)}\left[\frac{1}{(x_t - 1)}\ln x_t - \frac{1}{x_t}\right] +$$

$$- \frac{3}{4}\frac{x_t^2}{(x_t - 1)^2}\left[-\frac{2}{(x_t - 1)}\ln x_t + \frac{1}{x_t} + 1\right] =$$

$$= \frac{x_t}{2}F(x_t). \quad (13.28)$$

with $F(1) = \frac{3}{2}$, $F(+\infty) = \frac{1}{2}$.

Substituting [5] the values $M_W = 80.399$ GeV, $m_t = 172.9$ GeV, $x_t = 4.62$ gives:

$$F(x_t) = 1.093. \quad (13.29)$$

$E(x_c, x_c)$: It is obtained in the limit of (13.28) for $x \to 0$. We find:

$$E(x_c, x_c) = x_c. \quad (13.30)$$

$E(x_c, x_t)$:

$$E(x_c, x_t) = x_c x_t \int_0^\infty dt \left(1 + 2t + \frac{1}{4}t^2\right) \frac{1}{(t + x_c)(t + x_t)} \frac{1}{(t + 1)^2} =$$

$$= x_c x_t \left[\frac{1}{4}\mathcal{I}(x_c, x_t) + \frac{3}{2}\mathcal{I}_1(x_c, x_t, 1) - \frac{3}{4}\mathcal{I}_2(x_c, x_t, 1)\right]. \quad (13.31)$$

The integral has a logarithmic singularity for $x_c \to 0$ which is easily isolated, obtaining:

$$E(x_c, x_t) = x_c \ln \frac{1}{x_c}. \quad (13.32)$$

13.2 EFFECTIVE LAGRANGIAN AND VACUUM SATURATION

The amplitudes (13.16) and (13.18) can be obtained as matrix elements to perturbative first order in the effective Lagrangian $\mathcal{L}_{eff}$:

$$\mathcal{L}_{eff}(d\bar{s} \to \bar{d}s) = -\frac{G_F^2 M_W^2}{16\pi^2} \times \sum_{i,j=c,t} C_i C_j E(x_i, x_j) \times$$

$$\times \left[\bar{d}\gamma_\mu(1-\gamma_5)s\right] \left[\bar{d}\gamma^\mu(1-\gamma_5)s\right]. \qquad (13.33)$$

Now the symbols s and $\bar{d}$ denote the quark *fields*. If we take the matrix element of (13.33) between the states indicated in (13.3) and apply the contraction rules of the operators in the Lagrangian with the creation operators of the particles or antiparticles in the initial and final states, we find a factor two relative to the choice of the s field with which to annihilate the initial strange quark. There are two possibilities according to the choice of annihilating the anti-d of the initial state with the $\bar{d}$ field in the same covariant to which the s field belongs, or in the other. In the first case the result (13.16) is found for the diagram of Figure 13.1(a). In the second the result is (13.18) for the diagram of Figure 13.1(b).

The local operator

$$[\bar{d}(x)\gamma^\sigma(1-\gamma_5)s(x)] \qquad (13.34)$$

is able to annihilate an $s\bar{d}$ pair and create a $d\bar{s}$ pair. In the state with orbital angular momentum $L = 0$ and total spin $S = 0$, the quark-antiquark pair has the same quantum numbers as the pseudoscalar mesons $\bar{K}^0 = (s\bar{d})$ and $K^0 = (d\bar{s})$.

The operator (13.34) therefore has analogous properties to those of the $\bar{K}^0(x)$ field operator that is able to annihilate a $\bar{K}^0$ meson and create a K^0. With a certain approximation[3] we can identify the local operator (13.34) with the field of the $\bar{K}^0$, to within a suitable proportionality factor. Using isotopic spin symmetry, the factor is obtained from the amplitude describing the leptonic decay of the $K^- = (u\bar{s})$, which involves the matrix element of the charged current, the isospin partner of the neutral current appearing in (13.34). In turn, the kaon leptonic decay amplitude is parameterised in terms of a decay constant f_K. We write:

$$< 0|\bar{u}(0)\gamma^\mu(1-\gamma_5)s(0)|K^- >= if_K p^\mu < 0|K^-(0)|K^- > \qquad (13.35)$$

from which (the vector current has a vanishing matrix element because of parity):

$$\bar{u}(x)\gamma^\mu\gamma_5 s(x) = f_K \frac{\partial}{\partial x_\mu} K^-(x)$$

and by isotopic spin symmetry:

$$\bar{d}(x)\gamma^\mu\gamma_5 s(x) = f_K \frac{\partial}{\partial x_\mu} \bar{K}^0(x). \qquad (13.36)$$

---

[3]Which is better the smaller the mass of the meson.

Note. If we take the derivative of equation (13.36) and use the Dirac equations for the free quarks (see chapter 11), we obtain:

$$\partial_\mu(\bar{d}\gamma^\mu\gamma_5 s) = (m_s + m_d)(i\bar{d}\gamma_5 s) = -m_K^2\bar{K}^0. \tag{13.37}$$

The pseudoscalar density has the role of interpolating field of the $\bar{K}^0$. For the pseudoscalar density, it is easy to see (see [1], chapter 12) that, under CP transformation:

$$CP(i\bar{d}\gamma_5 s)(CP)^{-1} = -(i\bar{s}\gamma_5 d) \tag{13.38}$$

or

$$CP\bar{K}^0(CP)^{-1} = -K^0 \tag{13.39}$$

so justifying the convention chosen in chapter 12, equation (12.6).

Colour factor. To write correctly the effective Lagrangian in terms of the $\bar{K}^0$ and K^0 fields it is necessary to allow for the quark colour and the fact that the meson states are colour singlets. In the presence of colour, the operator (13.34) should be written as:

$$\bar{d}(x)\gamma^\sigma(1 - \gamma_5)s(x) = \sum_{a=1,3} \bar{d}^a(x)\gamma^\sigma(1 - \gamma_5)s_a(x) \tag{13.40}$$

and consequently:

$$[\bar{d}(x)\gamma^\sigma(1 - \gamma_5)s(x)][\bar{d}(x)\gamma^\sigma(1 - \gamma_5)s(x)] =$$
$$= \sum_{a,b=1,3} [\bar{d}^a(x)\gamma^\sigma(1 - \gamma_5)s_a(x)][\bar{d}^b(x)\gamma^\sigma(1 - \gamma_5)s_b(x)]. \tag{13.41}$$

To choose the pair of $\bar{d}$ and s fields to associate with the $\bar{K}^0$ meson we must once again make a double choice: (i) from which current to take the field of the relevant d antiquark, which gives us first a factor 2, and (ii) from which current to choose the field of the s quark. If we choose them in the same current, the fields are already in a colour singlet, but if we choose them in the other one we have the combination $\bar{d}^a s_b$ which must be projected onto the colour singlet and onto the correct spatial state. The second point presents no problem; we can use the Fierz transformation to show that, allowing for the anticommutation properties of the fermion fields:

$$[\bar{\psi}_1\gamma^\mu(1 - \gamma_5)\psi_2][\bar{\psi}_3\gamma^\mu(1 - \gamma_5)\psi_4] = +[\bar{\psi}_1\gamma^\mu(1 - \gamma_5)\psi_4][\bar{\psi}_3\gamma^\mu(1 - \gamma_5)\psi_2] \tag{13.42}$$

or:

$$[\bar{d}^a\gamma^\mu(1 - \gamma_5)s_a][\bar{d}^b\gamma^\mu(1 - \gamma_5)s_b] = [\bar{d}^a\gamma^\mu(1 - \gamma_5)s_b][\bar{d}^b\gamma^\mu(1 - \gamma_5)s_a]. \tag{13.43}$$

To associate the $\bar{K}^0$ meson with the first bilinear on the right hand side, we must project onto the colour singlets. This is done by using the completeness relation of the $SU(3)$ matrices, $\mathbf{1}$ and λ^A ($A = 1, \cdots, 8$), in analogy with the Fierz transformation. We can write:

$$\delta_d^a \delta_b^c = C_0 \delta_b^a \delta_d^c + C_8 \sum_A (\lambda^A)_b^a (\lambda^A)_d^c. \tag{13.44}$$

Saturating both sides with δ_a^b we find:

$$\delta_d^c = 3C_0 \delta_d^c + C_8 \sum_A Tr(\lambda^A)(\lambda^A)_d^c = 3C_0 \delta_d^c \tag{13.45}$$

given that the Gell–Mann matrices have zero trace. Therefore $C_0 = 1/3$. This result determines the colour factor between the operator which appears in the effective Lagrangian (13.33) and the operator $(K^0)^2$:

$$[\bar{d}(x)\gamma^\sigma(1 - \gamma_5)s(x)][\bar{d}(x)\gamma^\sigma(1 - \gamma_5)s(x)] =$$
$$= (1 + \frac{1}{3})(f_K \frac{\partial}{\partial x_\mu} \bar{K}^0 f_K \frac{\partial}{\partial x^\mu} \bar{K}^0) =$$
$$= \frac{4}{3}(f_K m_K)^2 (\bar{K}^0)^2 \tag{13.46}$$

and thus:

$$\mathcal{L}_{eff}(d\bar{s} \to \bar{d}s) = -\frac{G_F^2 M_W^2}{12\pi^2} \times$$
$$\times \sum_{i,j=c,t} C_i C_j E(x_i, x_j) \times m_K^2 f_K^2 \times (\bar{K}^0)^2 + \text{h.c.} + \cdots \tag{13.47}$$

where we have neglected terms in which the operator (13.47) connects the K meson to heavier intermediate states (cf. the following discussion on saturation of the vacuum).

Matrix element.   We are now able to calculate the matrix element of the effective Hamiltonian in the rest system of the K mesons, which gives us the off-diagonal element of the mass matrix M_{12}, equation (12.2):

$$M_{12}(\bar{K}^0 \to K^0) = < K^0 | \mathcal{H}_{eff} | \bar{K}^0 > = < K^0 | - \mathcal{L}_{eff} | \bar{K}^0 > =$$
$$= \frac{G_F^2 M_W^2}{12\pi^2} \times \sum_{i,j=c,t} C_i C_j E(x_i, x_j) \times m_K^2 f_K^2 < K^0 | (\bar{K}^0)^2 | \bar{K}^0 > =$$
$$= \frac{G_F^2 M_W^2}{12\pi^2} \times \sum_{i,j=c,t} C_i C_j E(x_i, x_j) \times m_K^2 f_K^2 \frac{2}{2m_K} =$$
$$= \frac{(G_F M_W^2)(G_F f_K^2)}{12\pi^2} \times \sum_{i,j=c,t} C_i C_j E(x_i, x_j) \times m_K. \tag{13.48}$$

Vacuum saturation. To appreciate the significance of (13.48), we consider the matrix element of the effective Lagrangian between K^0 and $\bar{K}^0$ states starting from:

$$< K^0|\mathcal{L}_{eff}(0)|\bar{K}^0 > = Const < K^0|[\bar{s}(0)\gamma^\sigma(1-\gamma_5)d(0)][\bar{s}(0)\gamma_\sigma(1-\gamma_5)d(0)]|\bar{K}^0 > \quad (13.49)$$

Inserting a complete set of states between the quark bilinears, we find:

$$<K^0|\mathcal{L}_{eff}|\bar{K}^0 > =$$

$$= Const \cdot 2 \sum_n < K^0|\bar{s}(0)\gamma^\sigma(1-\gamma_5)d(0)|n >< n|\bar{s}(0)\gamma^\sigma(1-\gamma_5)d(0)|\bar{K}^0 > =$$

$$= Const \cdot \frac{8}{3} \sum_n < K^0|\bar{s}(0)\gamma^\sigma(1-\gamma_5)d(0)|0 >< 0|\bar{s}(0)\gamma^\sigma(1-\gamma_5)d(0)|\bar{K}^0 > + \cdots$$

$$(13.50)$$

where we have isolated the contribution of the minimum energy state, the vacuum. If we now use equation (13.35), we obtain for the term with the vacuum the same matrix element that we would obtain with (13.48). Hence, to use (13.48) corresponds to assuming that the vacuum state makes the most important contribution to the sum (13.50). This approximation is called *vacuum saturation*.

To account for the intermediate states following the vacuum state, a parameter B_K (B-factor) is introduced, by writing the matrix element of the effective Lagrangian as:

$$< K^0|\mathcal{L}_{eff}(0)|\bar{K}^0 > = < K^0|\mathcal{L}_{eff}(0)|\bar{K}^0 > |_{vac.sat.} \times B_K =$$

$$= Const \cdot \frac{8}{3}m_K^2(f_K^2 B_K)| < 0|K^0|K^0 > |^2. \quad (13.51)$$

The calculation of the matrix element in (13.48) with lattice QCD leads to values of B_K close to unity [47], justifying qualitatively the vacuum saturation hypothesis.

13.3 $\bar{B}_D^0 \to B_D^0$, $\bar{B}_S^0 - B_S^0$ and $\bar{D}^0 \to D^0$ AMPLITUDES

B mesons.   B meson transitions correspond, in terms of quarks, to processes of the type:

$$b + \bar{q}' \to \bar{b} + q' \quad (13.52)$$

with $q' = d$ or s, described by the same diagrams as Figure 13.1 with obvious changes. The quarks exchanged in the internal lines are still u, c, t and we can use equation (13.48) with appropriate redefinition of the coefficients C_i.

D mesons.   In this case, the quarks exchanged in the internal lines are the *down*-type quarks, d, s and b. We immediately find for the effective Lagrangian and for the $D^0 - \bar{D}^0$ transition matrix element:

$$\mathcal{L}_{eff}(u\bar{c} \to c\bar{u}) = -\frac{G_F^2 M_W^2}{16\pi^2} \times$$

$$\times \sum_{i,j=s,b} C_i C_j E(x_i, x_j) [\bar{u}\gamma_\mu(1-\gamma_5)c] [\bar{u}\gamma^\mu(1-\gamma_5)c]; \quad (13.53)$$

$$M_{12}(\bar{D}^0 \to D^0) = \frac{(G_F M_W^2)(G_F f_D^2)}{12\pi^2} \times \sum_{i,j=s,b} C_i C_j E(x_i, x_j) \times m_D. \quad (13.54)$$

13.4   NUMERICAL ANALYSIS AND COMPARISON WITH DATA

The formulae for the CKM coefficients which appear in the K and B transition amplitudes and the corresponding numerical values are summarised in Tables 13.1 and 13.2. The values of the relevant Inami-Lim functions are shown below in Table 13.3.

The formulae in Table 13.1 are based on the simplified parameterisation of the CKM matrix shown in (10.19). This parameterisation is generally sufficient, except in the cc case in which the small imaginary part is not described. In this case, a more exact formula is obtained from (10.18):

$$C_c^2 = \left[-\lambda + A^2\lambda^2\left(1 - 2\rho + 2i\eta\right)\right]^2 \times \left[1 - \frac{1}{2}\lambda^2 - \frac{1}{8}\lambda^4(1 + 4A^2)\right]^2 =$$
$$= 5.07 \cdot 10^{-2} - i5.65 \cdot 10^{-5}. \tag{13.55}$$

To facilitate the comparison between the results with the two parameterisations, in Tables 13.1 and 13.2 we show in parentheses the values obtained with the most accurate parameterisation, equation (10.18).

Table 13.1   Expressions and values of the coefficients $C_iC_j = (U_{iq'}^*U_{iq})(U_{jq'}^*U_{jq})$ with $q = s, b;\ q' = d, s$, which determine the contributions of the quarks $i, j = c, t$ in the transitions with $\Delta F = 2$ of $K^0, B_d^0, \bar{B}_s^0$. For the CKM parameters cf. section 10.2. Expressions from equation (10.19); numerical values of C_iC_j from (10.19) or, in square brackets, from (10.18).

	$\bar{K}^0 \to K^0$	$\bar{B}_d^0 \to B_d^0$	$\bar{B}_s^0 \to B_s^0$
cc	$\lambda^2(1 - \lambda^2)$	$A^2\lambda^6$	$A^2\lambda^4(1 - \lambda^2)$
	$4.81\ 10^{-2}$	$8.51\ 10^{-5}$	$1.59\ 10^{-3}$
	$[5.07\ 10^{-2} - i5.65\ 10^{-5}]$	$[8.51\ 10^{-5} - i9.75\ 10^{-8}]$	$[1.59\ 10^{-3}]$
ct	$A^2\lambda^6(1 - \frac{1}{2}\lambda^2)(1 - \rho + i\eta)$	$-A^2\lambda^6(1 - \rho + i\eta)$	$-A^2\lambda^4(1 - \frac{1}{2}\lambda^2)$
	$(7.20 + i2.83)\ 10^{-5}$	$(-7.29 + i2.90)\ 10^{-5}$	$-1.64\ 10^{-3}$
	$[(7.04 + i2.83)\ 10^{-5}]$	$[(-7.39 + i2.78)\ 10^{-5}]$	$[-1.58\ 10^{-3} + i2.78\ 10^{-5}]$
tt	$A^4\lambda^{10}(1 - \rho + i\eta)^2$	$A^2\lambda^6(1 - \rho + i\eta)^2$	$A^2\lambda^4$
	$(9.11 + i8.46)\ 10^{-8}$	$(5.42 + i5.04)\ 10^{-5}$	$1.68\ 10^{-3}$
	$[(8.65 + i8.29)\ 10^{-8}]$	$[(5.34 + i4.77)\ 10^{-5}]$	$[1.56\ 10^{-3} - i5.51\ 10^{-5}]$

Decay constants of the pseudoscalar mesons.   In the vacuum saturation approximation, the matrix element of the effective Lagrangian requires the decay constant for the charged pseudoscalar mesons, f_P, defined in (13.36)

Table 13.2 Expressions and CKM coefficient values for the D^0 case, for $C_i C_j = (U_{ci} U_{ui}^*)(U_{cj} U_{uj}^*)$; $i, j = s, b$. Under each expression are the numerical values of the coefficients $C_i C_j$, equation (10.19) or (10.18).

	ss	sb	bb
$C_i C_j$	$\lambda^2(1-\lambda^2)$	$A^2\lambda^6(1-\frac{1}{2}\lambda^2)(\rho+i\eta)$	$A^4\lambda^{10}(\rho+i\eta)^2$
	$4.82\ 10^{-2}$	$(1.10-i2.83)\ 10^{-5}$	$(-1.41+i1.29)\ 10^{-8}$
	$[4.81\ 10^{-2}+i2.83\ 10^{-5}]$	$[(1.09+i2.83)\ 10^{-5}]$	$[(-1.41+i1.29)\ 10^{-8}]$

Table 13.3 Values of the Inami and Lim functions; $q_{1,2} = c$, t or s, b are the quarks exchanged in the loop. Values used for the masses (in GeV) are $m_c = 1.5$, $m_t = 173$, $m_s = 0.150$, $m_b = 5.0$.

	$E(x_1, x_1)$	$2E(x_1, x_2)$	$E(x_2, x_2)$
$q_{1,2} = c, t$	$3.48\ 10^{-4}$	$5.54\ 10^{-3}$	2.53
$q_{1,2} = s, b$	$3.48\ 10^{-6}$	$4.88\ 10^{-5}$	$3.87\ 10^{-3}$

for the K, with analogous extension to higher flavour mesons. To lowest order in the weak interactions, f_P is obtained from the decay rate of the process:

$$P^+ \to l^+ + \nu_l \qquad (13.56)$$

according to the formula:

$$\Gamma(P^+ \to l^+ + \nu_l) = \frac{G_F^2}{8\pi} f_P^2 m_l^2 m_P \left(1 - \frac{m_l^2}{m_P^2}\right) |U_{q_1 q_2}| \qquad (13.57)$$

where $U_{q_1 q_2}$ is the CKM matrix element for the annihilation of the quark-antiquark pair in the initial state. We list in Table 13.4 the values of the decay constants for π, K and D mesons [48]. Values of $f_{B_{d,s}}$ are obtained at present from lattice QCD calculations; cf. [49] and [51].

QCD corrections and B factors.   In the case of the parameters of the K^0, equation (13.48) is rewritten in the following way:

$$M_{12}(\bar{K}^0 \to K^0)|_{corr} = \frac{(G_F M_W^2)(G_F f_K^2)}{12\pi^2} \times$$
$$\times \left[\eta_1 C_c^2 E(x_c, x_c) + \eta_2 C_t^2 E(x_t, x_t) + 2\eta_3 C_c C_t E(x_c, x_t)\right] \times m_K \times B_K \qquad (13.58)$$

where the coefficients $\eta_{1,2,3}$ are the QCD corrections to the box diagrams

Table 13.4 Values of the decay constants for the pseudoscalar mesons, f_P (MeV), defined according to (13.36). For simplicity, we show only the larger error from those reported in [48]. The values of $B_{d,s}$ are obtained from lattice QCD calculations, which give directly the decay constant multiplied by the root of the B-factor, equation (13.51).

$f(\pi^+)$	$f(K^+)$	$f(D^+)$	$f(D_s^+)$	$\sqrt{B_d}f(B_d)$	$\xi = \frac{\sqrt{B_s}f(B_s)}{\sqrt{B_d}f(B_d)}$
130.41 ± 0.03	156.1 ± 0.8	206.7 ± 8.9	257.5 ± 6.1	230 ± 25	1.14 ± 0.04

and B_K the B factor. In [51] the references to the original articles can be found and a compilation of more recent values:

$$\eta_1 = 1.38; \quad \eta_2 = 0.574; \quad \eta_3 = 0.47; \quad B_K = 0.87; \qquad (13.59)$$

(for the next-to-leading order corrections, see [52]).

In the B meson case, the result is, as mentioned, dominated by the t quark and we write:

$$M_{12}(\bar{B}^0 \to B^0)|_{corr} = \frac{(G_F M_W^2)(G_F B_B f_B^2)}{12\pi^2} \eta_b C_t^2 E(x_t, x_t) \times m_B \quad (13.60)$$

where $\eta_b = 0.55$ represents the QCD correction and B_B the appropriate B-factor, Table 13.4.

Comparison with data. In the case of the K^0, the very small value of $Re(C_t)^2$ compared to $Re(C_c)^2$ should be noted. Despite the fact that $x_t \gg x_c$, the K^0–$\bar{K}^0$ mass difference is dominated by the exchange of the charm quark, as in the GIM theory, and the connection with the mass scale of the charm quark remains valid also in the CKM theory. Of course, top exchange dominates the CP violation parameter, ϵ, which would vanish with two generations only.

The CKM coefficients for B mesons are of the same order of magnitude and the contribution of the top quark dominates the real and imaginary parts of M_{12}.

The D^0 case is similar to that of the K^0. The (short distance?) sb term dominates the imaginary part of M_{12} while the (long distance) ss term dominates the real part, which however is essentially non-calculable due to strong interaction corrections, in particular the effect of $SU(3)$ breaking[4].

In conclusion, reliable predictions can be obtained for $ImM_{12}(K, B)$ and ReM_{12} for $B_{d,s}^0$, quantities which are dominated by short distance corrections with calculable QCD corrections (up to now, as far as next-to-next-to leading order terms in α_S). The case of $ReM_{12}(K^0)$ is at the edge of the trustworthy

---

[4]In the $m_s = m_d$ limit there is no mixing; for a recent discussion in the D^0 case see [53].

region. In what follows we concentrate on calculation of $M_{12}(K^0)$, real and imaginary parts of $M_{12}(K^0)$, and $|\Delta M|$ for B_d^0 and B_s^0.

We list in the first line of Table 13.5 the results of calculations of the relevant parameters for K^0, B_d^0 and B_s^0, obtained from the box diagram amplitudes and the numerical values given in Tables 13.1, 13.3 and 13.4. We recall that $|\Delta M(B)| = 2|M_{12}|$; cf. equation (12.36).

The calculated values, while generally of the correct order, still differ considerably from the observed values reported in the third line. The agreement improves substantially if QCD corrections and B factors obtained from lattice QCD are introduced, as shown in the second line.

Table 13.5 Experimental values of selected CP-violating and CP-conserving observables compared with predictions from the box-diagram amplitudes without or with QCD corrections, first and second line respectively. Masses in MeV.

	$|\epsilon_K|$	Δm_K	$|\Delta M(B_d^0)|$	$|\Delta M(B_s^0)|$	$|\Delta M(D^0)|$
diag. box	$6.34\ 10^{-3}$	$3.12\ 10^{-12}$	$7.51\ 10^{-10}$	$294\ 10^{-10}$	$2.0\ 10^{-13} \cdot \left(\frac{m_s}{0.15GeV}\right)^2$
QCD & B-fact.	$2.65\ 10^{-3}$	$3.85\ 10^{-12}$	$4.13\ 10^{-10}$	$119\ 10^{-10}$	??
expt.	$2.228\ 10^{-3}$	$3.483\ 10^{-12}$	$3.34\ 10^{-10}$	$117.0\ 10^{-10}$	$(1.57 \pm 0.39)\ 10^{-11}$

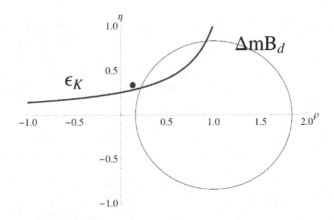

Figure 13.2 Determination of the ρ and η coefficients of the CKM matrix obtained from the values of $|\epsilon_K|$ and $|\Delta M(B_d^0)|$ using the theoretical formula in the text. The point in the figure is the value reported by the PDG, equation (10.20).

Considering ρ and η as free parameters, we can trace the curves in the $\rho - \eta$ plane which correspond to the experimental values of ϵ_K and $\Delta M(B_d^0)$ and which identify, by their intersection, the values of the coefficients ρ and η of the CKM matrix. The result is shown in Figure 13.2 and is in good agreement with the values from the Particle Data Group, equation (10.20) represented by the point shown in Figure 13.2 and obtained by using available information on weak transitions, with and without CP violation.

THE SEARCH FOR THE HIGGS BOSON

CONTENTS

We summarise first the relevant formulae for the electroweak interactions of the doublet of the Higgs field (see section 6.2):

$$\phi = \begin{pmatrix} \phi^+ \\ \phi^0 \end{pmatrix}_{Y=+1}. \tag{14.1}$$

We write the Lagrangian for the gauge fields and the scalar doublet:

$$\mathcal{L}_{tot} = \mathcal{L}_W + \mathcal{L}_{\phi W} \tag{14.2}$$

where

$$\mathcal{L}_W = \frac{1}{4}[\mathbf{W}_{\mu\nu}\mathbf{W}^{\mu\nu} + B_{\mu\nu}B^{\mu\nu}] \tag{14.3}$$

while

$$\mathcal{L}_{\phi W} = (D_\mu\phi)^\dagger (D^\mu\phi) - V(\phi);$$
$$V(\phi) = \mu^2\phi^\dagger\phi + \lambda(\phi^\dagger\phi)^2;$$
$$\mu^2 < 0; \tag{14.4}$$

with the definition of the covariant derivative:

$$D_\mu\phi = [\partial_\mu + ig\mathbf{W}_\mu \cdot \frac{\tau}{2} + ig'(+\frac{1}{2})B_\mu]\phi. \tag{14.5}$$

14.1   INTERACTIONS OF THE HIGGS BOSON

We can rewrite the potential of the scalar field as:

$$V = \lambda(\phi^\dagger\phi + \frac{\mu^2}{2\lambda})^2 + constant. \tag{14.6}$$

If we omit the non-essential constant, the minimum of the potential is zero and occurs for:

$$\phi^\dagger\phi = \eta^2 = -\frac{\mu^2}{2\lambda} > 0 \tag{14.7}$$

($\mu^2 < 0$). In the unitary gauge we set:

$$\phi(x) = \begin{pmatrix} 0 \\ \eta + \frac{\sigma(x)}{\sqrt{2}} \end{pmatrix} \tag{14.8}$$

and expand the Lagrangian (14.2) in powers of the fields.

The quadratic terms give the mass of the Higgs boson and the masses of the vector bosons, which are diagonalised as explained in chapter 6. The higher order terms describe interactions of the vector bosons with each other and of the field σ with the vector bosons.

The value of M_σ.   For the mass of the Higgs boson we find:

$$M_\sigma^2 = 4\lambda^2\eta^2. \tag{14.9}$$

The η value of 174 GeV is fixed by the Fermi constant (see (6.22)), and the mass depends on the interaction constant which appears in the potential:

$$M_\sigma = \sqrt{\lambda}\ 340 \text{ GeV}. \tag{14.10}$$

Interactions with the vector bosons.   The vacuum expectation value of the field ϕ generates the mass term of the vector fields which, expressed in terms of the physical fields $W^\pm$ and Z, is written:

$$\mathcal{L}_{W-mass} = M_W^2 W^\mu W_\mu^\dagger + \frac{1}{2}M_Z^2 Z^\mu Z_\mu. \tag{14.11}$$

We can obtain the complete Lagrangian simply by substituting into the mass formula:

$$\eta \to \eta + \sigma/\sqrt{2}. \tag{14.12}$$

We find:

$$\mathcal{L}_{W-\sigma} = \mathcal{L}_{W-mass} + (\sqrt{2}\frac{\sigma}{\eta} + \frac{\sigma^2}{2\eta^2})(M_W^2 W^\mu W_\mu^\dagger + \frac{1}{2}M_Z^2 Z^\mu Z_\mu). \tag{14.13}$$

The Lagrangian describes the σ–VV couplings, which are typical of spontaneous symmetry breaking (the photon, which remains massless, does not couple directly to the Higgs), and the quartic couplings, σ–σ–VV. These are simply the reflection of the 'seagull' term which originates in the square of the covariant derivative. This coupling, typical of the interaction of a scalar field with the vector field, does not depend on the symmetry breaking. The explicit factor $1/\eta^2$ is cancelled by the factor η^2 in the squared masses of the vector bosons and the result is proportional to the dimensionless constants g^2 or g'^2.

Using (6.22), we write the σ–VV interaction as:

$$\mathcal{L}_{\sigma-VV} = 2(\sqrt{2}G_F)^{1/2} \, \sigma \, (M_W^2 W^\mu W_\mu^\dagger + \frac{1}{2}M_Z^2 Z^\mu Z_\mu). \tag{14.14}$$

Interactions with the fermions. The mass term of the electron arises from the Lagrangian through interactions between the Higgs field and the electron; see (6.30). With the substitution (14.12) we obtain the σ–ee Yukawa interaction:

$$\mathcal{L}_{\sigma-ee} = \mathcal{L}_{e-mass} + g_e \, \frac{\sigma}{\sqrt{2}} \, \bar{e}e = \mathcal{L}_{e-mass} + \frac{m_e}{\sqrt{2}\eta} \, \sigma\bar{e}e. \tag{14.15}$$

If we use equation (6.32), the σ–e Yukawa interaction is written, finally:

$$\mathcal{L}_e^Y = \left(\sqrt{2}G_F\right)^{1/2} m_e \, \sigma\bar{e}e. \tag{14.16}$$

A similar result is obtained for the coupling of the other fermions with the overall result:

$$\mathcal{L}^Y = \left(\sqrt{2}G_F\right)^{1/2} \sum_f m_f \, \sigma \, \bar{f}f \tag{14.17}$$

where the sum is over all the fermions of the theory, quarks and leptons, and m_f is the physical mass of each fermion.

14.1.1   Decay width

From the preceding formulae it follows that the $\sigma \to f\bar{f}$ and $\sigma \to VV$ decay widths are proportional to the square of the fermion mass, or to the fourth power of $M_{W,Z}$, with a proportionality constant fixed by the Fermi constant. This rule reflects the fact that fermions and intermediate bosons acquire their mass via the interaction with the Higgs field and it is the characteristic *signature* of the Higgs boson.

Among fermionic decays, the dominant channel is that of the heaviest fermion, the top quark, if the mass allows it (threshold at $M_\sigma = 2m_t \simeq 348$ GeV), or the beauty quark (threshold at $M_\sigma = 2m_b \simeq 10$ GeV).

Among other important, or even dominant, channels are the W^+W^- (threshold at $M_\sigma = 2M_W \simeq 160$ GeV) and ZZ (threshold at $M_\sigma = 2M_Z \simeq 180$ GeV) decays. These decays are also important below threshold, with one of the intermediate bosons being virtual, in which case it resembles a β decay:

$$\sigma \to W^+ \, l^- \, \bar{\nu}_l;$$
$$\sigma \to W^- \, l^+ \, \nu_l;$$
$$\sigma \to Z \, l^+ l^-. \tag{14.18}$$

where l is one of the charged leptons and ν_l is the corresponding neutrino. The real intermediate bosons decay in their turn into hadrons or lepton pairs. The most precise signature is obtained in the cases in which $Z \to l'^+ l'^-$, leading to the final state with two pairs of charged leptons:

$$\sigma \to l^+ l^- l'^+ l'^-$$
$$M(4 \text{ lepton}) = M_\sigma. \tag{14.19}$$

Although rather rare, this decay can be distinguished from background by a peak in the mass distribution of the lepton system, at the mass M_σ.

Figure 14.1 shows the values of the branching ratios for decays into different channels as a function of the mass of the Higgs boson, σ.

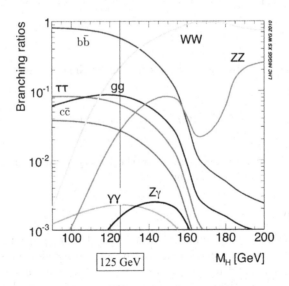

Figure 14.1 Branching ratios, σ, for decays of the Higgs boson as a function of its mass. Figure by LHC Higgs Working Group, CERN 2011. The vertical line at 125 GeV is a guide to read the predicted values for the particle observed at CERN.

14.1.2 Calculations

We illustrate the calculation of the decay width of the Higgs boson in several channels important for its identification. Where possible the calculations are carried out to lowest order perturbation theory, using the values of the quark and charged lepton masses given in section 10.3.

Table 14.1 compares the lowest order results with those of more advanced calculations, in which all the higher order corrections due to strong (QCD) interactions, in the quark case, and electroweak interactions, in the Standard Theory, are taken into account. These latest values are obtained with the HDECAY [54] code. In decays into quark-antiquark pairs, the QCD corrections are essential to obtain a quantitative prediction.

In numerical calculations we assume a value of the Higgs mass, $M_\sigma = 125$ GeV, which corresponds to the mass of the particle observed in the most recent LHC (Large Hadron Collider) experiments at CERN, as a possible candidate for the Higgs boson.

Table 14.1 Theoretical predictions for the decay widths of the Higgs boson to lowest order (second column) and with higher order QCD and electroweak corrections included (third column). In the fourth column, the radiatively corrected branching ratios are reported.

Decay mode	Γ(MeV) (lowest order)	Γ(MeV) (all corrections)	Branching fraction (%) (corrected)
$b\bar{b}$	4.92	2.36	56.6
$\tau\bar{\tau}$	0.259	0.259	6.2
$c\bar{c}$	0.50	0.12	2.9
$W\,W^*$	0.779	0.941	22.5
$Z\,Z^*$	0.0839	0.119	2.9
$g\,g$	0.22	0.35	8.5
$\gamma\,\gamma$	0.0091	0.0096	0.23
Total	6.77	4.17	99.8

$\Gamma(\sigma \to f\bar{f})$. We start from the Feynman matrix element and we calculate its square, summed over the final degrees of freedom (spin and, in the case of quarks, colour):

$$\mathcal{M} = \left(\sqrt{2}G_F\right)^{1/2} m_f \bar{u}_f v_f$$

$$\sum_{\text{spin\&col}} |\mathcal{M}|^2 = \left(\sqrt{2}G_F m_f^2\right) N_c \text{Tr} \frac{[(\not{p} + m_f)(\not{q} - m_f)]}{4m_f^2} =$$

$$= \frac{G_F m_f^2}{\sqrt{2}} N_c \left(\frac{M_\sigma^2 - 4m_f^2}{m_f^2}\right) \tag{14.20}$$

where $N_c = 1$ or 3 for the leptons and quarks, respectively. The decay width is obtained from the formula (see [1], section 11.3):

$$\Gamma(f\bar{f}) = \frac{G_F m_f^2}{2\sqrt{2}M_\sigma} N_c \int \frac{d^3p\,d^3q}{(2\pi)^6} (2\pi)^4 \delta^{(4)}(P_\sigma - p - q) \frac{m_f^2}{E_f E_{\bar{f}}} \left(\frac{M_\sigma^2 - 4m_f^2}{m_f^2}\right)$$

$$= \frac{G_F m_f^2}{2\sqrt{2}} N_c M_\sigma \left(1 - \frac{4m_f^2}{M_\sigma^2}\right) \int \frac{d^3p\,d^3q}{4\pi^2\,E_f E_{\bar{f}}} \delta^{(4)}(P_\sigma - p - q). \tag{14.21}$$

The integral which appears in the previous formula is Lorentz invariant. In the σ rest system the final particles have equal and opposite momenta and the same energy.

$$J = \int \frac{d^3p\,d^3q}{E_f^2} \delta^{(3)}(\mathbf{p} + \mathbf{q})\delta(M_\sigma - 2E_f) =$$

$$= \int 4\pi \frac{pE_f dE_f}{E_f^2} \delta(M_\sigma - 2E_f) = 2\pi \frac{p}{E_f} = 2\pi \left(1 - \frac{4m_f^2}{M_\sigma^2}\right)^{1/2}. \tag{14.22}$$

Putting everything together, finally we find:

$$\Gamma(f\bar{f}) = \frac{G_F m_f^2}{4\sqrt{2}\pi} N_c M_\sigma \left(1 - \frac{4m_f^2}{M_\sigma^2}\right)^{3/2}. \tag{14.23}$$

In the case of the τ lepton ($N_c = 1$), the formula (14.23) leads to the value given in the first column of Table 14.1, in excellent agreement with the value obtained by including the higher order electroweak corrections (second column).

This is not the case for the beauty or charm quark ($N_c = 3$), for which the QCD corrections are significant and must be included to obtain a useful comparison with experiment.

Note 1. The main effect of the QCD corrections is to make the value of the quark mass dependent on the momentum scale on which the process occurs.

In going from momenta of order of the quark masses themselves (the spectroscopic value) to a scale of order 10^2 GeV, the mass of the beauty quark is renormalised by a factor estimated to be about 1/3, which leads to the result of the second column of Table 14.1. The same considerations hold for the charm quark, for which, however, the correction calculation is less reliable because it involves relatively small momentum scales, for which QCD is not fully in the perturbative regime.

Note 2. The quantity raised to the power $\frac{3}{2}$ in (14.23) is the squared velocity of the fermion, β^2, in the σ rest system. To form a state with $J^{PC} = 0^{++}$, allowing for the opposite intrinsic parity of fermion and antifermion, the final particle pair must be in a state with orbital angular momentum $L = 1$. This explains the presence in (14.23) of the factor $\beta^3 = \beta^{2L+1}$, which represents the familiar centrifugal barrier factor associated with the P-wave.

$\Gamma(\sigma \to WW^*)$. We begin by calculating the width of the process (Figure 14.2):

$$\sigma \to W^+ e^- \bar{\nu}_e \tag{14.24}$$

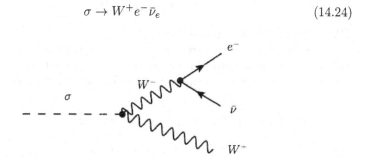

Figure 14.2 Higgs boson decay into a real and a virtual W.

The invariant matrix element is obtained from the effective Lagrangian (14.14) and is written:

$$\mathcal{M} = 2(\sqrt{2}G_F)^{1/2} M_W^2 \epsilon_\mu(k) \times$$
$$\times \frac{g}{2\sqrt{2}} \left(g_{\mu\nu} - \frac{q_\mu q_\nu}{M_W^2} \right) \frac{1}{q^2 - M_W^2} \bar{u}(e)\gamma^\nu(1 - \gamma_5)v(\nu) \tag{14.25}$$

where ϵ_μ is the polarisation vector of the W^+, k its momentum and $q = e + \nu$ (we denote the momentum of the electron and antineutrino by the name of the particle). The width is obtained from the formula:

$$\Gamma(\sigma \to W^+ e^- \bar{\nu}_e) = \frac{1}{2M_\sigma} \int \frac{d^3k \, d^3e \, d^3\nu}{2E_W (2\pi)^9} (2\pi)^4 \delta^{(4)}(P_\sigma - k - e - \nu) \times$$

$$\times \frac{m_e m_\nu}{E_e E_\nu} \sum_{spin} |\mathcal{M}|^2; \qquad (14.26)$$

$$\frac{m_e m_\nu}{E_e E_\nu} \sum_{spin} |\mathcal{M}|^2 =$$

$$= \frac{\lambda^2}{(q^2 - M_W^2)^2} \frac{Tr \left[\not{e} \gamma_\mu (1 - \gamma_5) \not{\psi} \gamma_\nu (1 - \gamma_5) \right]}{4 E_e E_\nu} \left(-g^{\mu\nu} + \frac{k^\mu k^\nu}{M_W^2} \right) =$$

$$= \frac{2\lambda^2}{(q^2 - M_W^2)^2} \frac{[2e_\mu \nu_\nu - g_{\mu\nu}(e \cdot \nu)]}{E_e E_\nu} \left(-g^{\mu\nu} + \frac{k^\mu k^\nu}{M_W^2} \right) \qquad (14.27)$$

where $(e \cdot \nu) = e_\mu \nu^\nu$ and:

$$\lambda^2 = 4(G_F M_W^2)^2 M_W^2. \qquad (14.28)$$

(We have used (4.33) to eliminate g.)

First, it is useful to carry out the integration over the momenta of the electron and the neutrino, which leads to a covariant result. To do this, we insert at the beginning of the right hand side of (14.26) a factor:

$$1 = \int d\mu^2 \int d^4q \, \delta^{(4)}(q - e - \nu) \delta(q^2 - \mu^2) =$$

$$= \int d\mu^2 \int \frac{d^3q}{2E_q} \delta^{(4)}(q - e - \nu) \qquad (14.29)$$

and we obtain:

$$\Gamma(\sigma \to W^+ e^- \bar{\nu}_e) = \frac{2\lambda^2}{2M_\sigma (2\pi)^5} \int \frac{d\mu^2}{(\mu^2 - M_W^2)^2} \times$$

$$\times \int \frac{d^3k \, d^3q}{4 E_W E_q} \delta^{(4)}(P_\sigma - k - q) \left(-g^{\mu\nu} + \frac{k^\mu k^\nu}{M_W^2} \right) L^{\mu\nu}(q) \qquad (14.30)$$

with

$$L^{\mu\nu} = \int \frac{d^3e \, d^3\nu}{E_e E_\nu} \delta^{(4)}(q - e - \nu) \left[2e_\mu \nu_\nu - g_{\mu\nu}(e \cdot \nu) \right]. \qquad (14.31)$$

In the limit in which we set to zero the mass of the electron and the neutrino, $q_\mu L^{\mu\nu} = 0$, from which

$$L^{\mu\nu} = A(q^2 g^{\mu\nu} - q^\mu q^\nu). \qquad (14.32)$$

Multiplying by $g_{\mu\nu}$ and carrying out the integration, it is easily found that:

$$A = -\frac{2\pi}{3}. \tag{14.33}$$

Substituting into (14.30) and carrying out the relevant integration for the decay into two particles of mass M_W and μ, we find:

$$\Gamma(\sigma \to W^+ e^- \bar{\nu}_e) = \frac{(G_F M_W^2)^2 M_W^2}{6M_\sigma \pi^3} \times$$
$$\times \int \frac{d\mu^2}{(\mu^2 - M_W^2)^2} \left[\mu^2 + \frac{(M_\sigma^2 - M_W^2 - \mu^2)}{8M_W^2} \right] \frac{P}{M_\sigma}; \tag{14.34}$$

$$P = \frac{\sqrt{\Delta}}{2M_\sigma};$$
$$\Delta(M_\sigma^2, M_W^2, \mu^2) = M_\sigma^4 + M_W^4 + \mu^4 - 2M_\sigma^2 M_W^2 - 2M_\sigma^2 \mu^2 - 2M_W^2 \mu^2. \tag{14.35}$$

P is the decay momentum in the σ rest system and Δ the familiar triangle function which determines the phase space of the two-body decay. The range of integration is from zero to $\mu^2 = (M_\sigma - M_W)^2$.

We normalise masses and widths to M_σ, introducing the variables:

$$z = \frac{M_W}{M_\sigma}; \quad u = \frac{\mu^2}{M_\sigma^2} \tag{14.36}$$

and find:

$$\Gamma(\sigma \to W^+ e^- \bar{\nu}_e) = M_\sigma \frac{(G_F M_W^2)^2}{12\pi^3} J(z);$$
$$J(z) = z^2 \int_0^{(1-z)^2} \frac{du}{(u - z^2)^2} \left[u - \frac{(1 - z^2 - u)^2}{8z^2} \right] \sqrt{\Delta(1, z^2, u)}. \tag{14.37}$$

To obtain the inclusive width with either of the two Ws which decays in all possible ways, we must multiply by a factor two and by the ratio between the total width of the W and the partial width into the $e\bar{\nu}$ channel. With this width set to one, the total width is equal to the total number of decay channels, which is:

$$N_{tot} = 3 + 2 \times 3 = 9 \tag{14.38}$$

(three lepton doublets and two active quark doublets, each in three colours). Hence, in total, we must multiply by 18 and we find, finally:

$$\Gamma(\sigma \to WW^*) = M_\sigma \frac{3(G_F M_W^2)^2}{2\pi^3} J(\frac{M_W}{M_\sigma}). \tag{14.39}$$

Numerically, for $M_\sigma = 125$ GeV, we find:

$$\boxed{\begin{array}{c} M_\sigma \frac{3(G_F M_W^2)^2}{2\pi^3} = 34.3 \text{ MeV} \\[2mm] J(\frac{M_W}{M_\sigma}) = 0.0227 \end{array}} \qquad (14.40)$$

from which the width given in the second column of Table 14.1 is obtained.

$\Gamma(\sigma \to ZZ^*)$. We write the single inclusive channel as:

$$\sigma \to Z + f + \bar{f}. \qquad (14.41)$$

The corresponding matrix element is (cf. equations (14.14) and (4.31)):

$$\mathcal{M}_Z = 2(\sqrt{2}G_F)^{1/2} M_Z^2 \epsilon_\mu(k) \times$$
$$\times \frac{g}{2\cos\theta_W}\left(g_{\mu\nu} - \frac{q_\mu q_\nu}{M_Z^2}\right)\frac{1}{q^2 - M_Z^2} \times$$
$$\times \bar{u}_f(p')\gamma^\nu\left[g_L(1 - \gamma_5) + g_R(1 + \gamma_5)\right]u_f(p) \qquad (14.42)$$

where

$$g_L = \frac{1}{2}\tau^3 - Q\sin^2\theta_W; \qquad g_R = -Q\sin^2\theta_W. \qquad (14.43)$$

If we consider the squared matrix element and compare it with the expression in (14.27), we see that:

$$\sum_{spin \ \& \ colour} |\mathcal{M}_Z|^2 = 2(g_L^2 + g_R^2) \times \sum_{spin \ \& \ colour} |\mathcal{M}|^2_{M_W \to M_Z}. \qquad (14.44)$$

If we want the inclusive width into the ZZ^* channel we must sum over all available fermion pairs (without the factor two which we put for the Ws because the Zs are indistinguishable). Explicitly:

$$\sum_f (g_L^2 + g_R^2) =$$
$$= 3(g_L^2 + g_R^2)_\nu + 3(g_L^2 + g_R^2)_e + 2 \times 3(g_L^2 + g_R^2)_{up} + 3 \times 3(g_L^2 + g_R^2)_{down} =$$
$$= \frac{21}{4} - 10\sin^2\theta_W + \frac{40}{3}\sin^4\theta_W \qquad (14.45)$$

(cf. the values of $g_{L,R}$ in Table 10.1).

Finally comparing with (14.39), we find:

$$\Gamma(\sigma \to ZZ^*) = M_\sigma\frac{3(G_F M_Z^2)^2}{2\pi^3} g_Z(\sin^2\theta_W) J(\frac{M_Z}{M_\sigma});$$
$$g_Z(\sin^2\theta_W) = \frac{7}{12} - \frac{10}{9}\sin^2\theta_W + \frac{40}{27}\sin^4\theta_W. \qquad (14.46)$$

Numerically, for $M_\sigma = 125$ GeV and using the value of $\sin^2\theta_W(M_Z^2)$, equation (7.28), we find:

$$\left[\begin{array}{c} M_\sigma \frac{3(G_F M_Z^2)^2}{2\pi^3} g_Z = 23.0 \text{ MeV} \\[2mm] J(\frac{M_Z}{M_\sigma}) = 0.00366 \end{array} \right] \tag{14.47}$$

and the width given in the second column of Table 14.1 is obtained.

Note. The analytic expression of the integral $J(z)$ in (14.37) is found in the literature, see for example [55] and [56]. For $1 > z^2 > 1/4$:

$$8J(z) = -(1 - z^2)(\frac{47}{2}z^2 - \frac{13}{2} + \frac{1}{z^2})-$$
$$- 3(1 - 6z^2 + 4z^4)\log(z)+$$
$$+ 3\frac{(1 - 8z^2 + 20z^4)}{\sqrt{4z^2 - 1}}\arccos(\frac{3z^2 - 1}{2z^3}). \tag{14.48}$$

14.1.3 γ–γ and gluon–gluon coupling

The coupling of σ to two photons occurs through the emission and reabsorption of charged particles. In the electroweak theory we are considering, the dominant terms are due to the top quark and the charged intermediate boson, Figure 14.3. Quarks lighter than t give a negligible contribution.

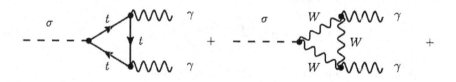

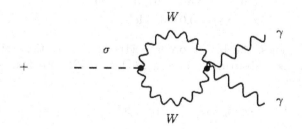

Figure 14.3 Leading Feynman diagrams for the Higgs boson decay into two photons.

$\Gamma(\sigma \to \gamma\gamma)$. If the σ mass is less than the threshold for production of real W or t pairs, the amplitude corresponding to the diagrams in Figure 14.3 can be represented as the matrix element of an effective Lagrangian which, taking into account the gauge invariance of the electromagnetic interactions, can be written succinctly in the form:

$$\mathcal{L}_{eff} = \frac{\alpha(\sqrt{2}G_F)^{1/2}}{8\pi} F(\sigma \, F_{\mu\nu} F^{\mu\nu}). \tag{14.49}$$

The calculation of the constant F has been carried out in [57, 58] and recently confirmed, see [59, 60]. We give the final result:

$$F = A_1(\tau_W) + N_c Q_t^2 A_{1/2}(\tau_t);$$
$$\tau_i = \frac{4m_i^2}{M_\sigma^2}. \tag{14.50}$$

For $\tau_i > 1$, i.e. σ below threshold, one finds:

$$A_1(x) = -x \left[\frac{2}{x} + 3 + 3(2-x)\arcsin^2(1/\sqrt{x}) \right];$$
$$A_{1/2} = 2x \left[1 + (1-x)\arcsin^2(1/\sqrt{x}) \right]. \tag{14.51}$$

For $\tau_i < 1$, i.e. σ above the pair production threshold:

$$\arcsin^2(1/\sqrt{x}) \to -\frac{1}{4} \left[\log\left(\frac{1+\sqrt{1-x}}{1-\sqrt{1-x}} \right) - i\pi \right]^2 \quad (x < 1). \tag{14.52}$$

From (14.49) we easily obtain[1]:

$$\Gamma(\sigma \to \gamma\gamma) = \frac{\alpha^2}{(4\pi)^2} \frac{G_F M_\sigma^2 |F|^2}{8\sqrt{2}\pi} M_\sigma. \tag{14.53}$$

Numerically, with $N_c = 3$, $Q_t = 2/3$, one finds:

$$A_1(\tau_W) = -8.32; \quad A_{1/2}(\tau_t) = 1.38;$$
$$\Gamma(\sigma \to \gamma\gamma) = 0.00909 \text{ MeV}.$$

As Table 14.1 shows, the higher order electroweak corrections are significant. Actually, just substituting $\alpha \to \alpha(M_Z^2) \simeq 1/128$ into (14.53), see equation (8.29), gives:

$$\Gamma(\sigma \to \gamma\gamma)_{\alpha(M_Z^2)} = 0.0104 \text{ MeV}. \tag{14.54}$$

---

[1]The two photon phase space is divided by two because of the identical particles.

$\Gamma(\sigma \to gg)$. The coupling of σ to two gluons occurs through the first diagram of Figure 14.3, with the substitution of a gluon of a given type for the photon. There are 8 different types of gluon, each associated with a colour $SU(3)$ matrix $\lambda^A, A = 1, \cdots, 8$. The QED interaction is replaced by the QCD interaction with the substitution (chapter 8):

$$\bar{q}\gamma^\mu eQA_\mu q \to \sum_{A=1,8} \bar{q}\gamma^\mu g_S \frac{\lambda^A}{2} A_\mu^A q \qquad (14.55)$$

where g_S is the QCD coupling constant and $\alpha_S = g_S^2/(4\pi) \simeq 0.119$ is the analogue of $\alpha \simeq 1/137$. Consequently, the factor $\alpha N_c Q_t^2$ in (14.50) is replaced by (cf. equation (2.58)):

$$\alpha N_c Q_t^2 \to \alpha_S \text{Tr}(\frac{\lambda^A \lambda^B}{4}) = \frac{\alpha_S}{2} \delta^{AB}. \qquad (14.56)$$

The two gluons must be of the same type, for example $g^1 g^1$, and the width into one of these states is found from that for two photons by the substitution:

$$(\alpha N_c Q_t^2)^2 \to \frac{\alpha_S^2}{4}. \qquad (14.57)$$

Since there are 8 different channels in total, we find, starting from (14.53),

$$\Gamma(\sigma \to \text{gg}) = \frac{\alpha_S^2}{(4\pi)^2} \frac{G_F M_\sigma^2}{4\sqrt{2}\pi} M_\sigma |A_{1/2}(\tau_t)|^2 = 0.218 \text{ MeV}. \qquad (14.58)$$

Note. In contrast to photons, gluons in the $\sigma \to gg$ decay are not directly observable, but materialise as two jets of hadronic particles. At the LHC, the background of hadron jets produced in proton–proton collisions is overwhelming and, unlike the two photon decay channel, the two gluon channel is not, at least at present, used for Higgs detection. The σgg coupling produced by the top quark is however very important in σ production at the LHC, which predominantly takes place through the gluon–gluon fusion mechanism, $gg \to \sigma$.

14.2 SEARCHES AT e^+e^- and PROTON–PROTON COLLIDERS

An important contribution to the search for the Higgs boson has originated from experiments at the CERN electron-positron collider, LEP. Not only have these experiments set a firm lower limit of 114 GeV on the σ mass but, together with the searches carried out at the proton-antiproton collider at Fermilab, USA, have allowed to develop methods for the Higgs boson searches in the presence of significant backgrounds which are of great importance for present experiments at the CERN proton–proton collider, the LHC.

σ production in electron-positron reactions takes place predominantly through the *Higgs bremsstrahlung* reaction illustrated in Figure 14.4. At an energy of 104 GeV per beam, reached by LEP at the end of 2000, the reaction in Figure 14.4 theoretically allows production of a σ up to a mass of 118 GeV, at which value the cross section vanishes.

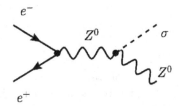

Figure 14.4 The dominant process for Higgs boson production in electron-positron collisions.

The cross sections of processes observed at LEP are shown in Figure 14.5, taken from [61]. The cross section for the process of Figure 14.4 corresponds to the dashed line in the lower right hand corner. The scale of the cross section compared to those of the dominant processes illustrates well the challenge of the search for the Higgs boson.

With the final LEP luminosity of $2 \cdot 10^{31}$ cm^{-2}s^{-1}, the four LEP experiments set a lower limit on the Higgs boson mass of 114 GeV.

σ production in hadronic collisions is based on several different processes, in the first place *gluon–gluon fusion*, Figure 14.6. This process is made possible by the high probability of finding a gluon in a high energy proton, section 8.4. The triangular vertex in the figure represents the σ–gg coupling via the t quark loop, discussed in section 14.1. The corresponding process is calculated starting from the gluon probability density functions, determined in their turn starting from cross sections for inelastic scattering of electrons and neutrinos and corresponds to the inclusive reaction:

$$p + p \rightarrow \sigma + \cdots \tag{14.59}$$

If we require two jets in the final state, gluon fusion comes about through the diagrams of Figure 14.7 (a)-(c) which correspond to the quasi-inclusive processes:

$$p + p \rightarrow \sigma + 2 \, \text{jet} + \cdots \tag{14.60}$$

The triangular vertex again represents the σ–gg coupling mediated by the t quark.

The requirement of two accompanying high momentum-transfer jets reduces the contributions of processes (a)–(c) to a level comparable with the vector boson fusion process (*VV* fusion), Figure 14.7 (d). In Table 14.2 we give the results of a recent calculation [62] which clearly shows the effect on

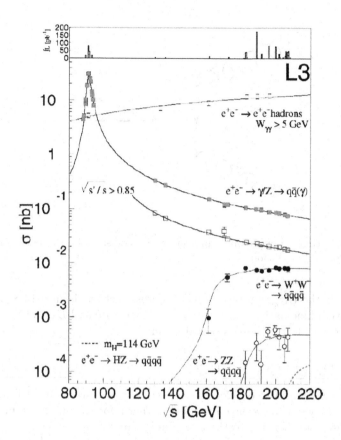

Figure 14.5 Cross sections for reactions observed at LEP by the L3 collaboration, as a function of centre of mass energy. For a discussion of the reactions and results from LEP, see [61]. In the lower right hand corner the cross section for reaction (14.4) is shown, dashed line, for $M_\sigma = 114$ GeV.)

the evaluation of the VV fusion process when including the presence of two high transverse momentum jets, as required by CMS and ATLAS.

Finally, Figure 14.8 shows the results of a precision calculation of the cross sections for production of a Higgs boson at the LHC at a centre of mass energy of 8 TeV, on the part of the LHC Higgs cross section Working Group [63].

It should be noted, with that a total cross section of around 20 pb (or $20 \cdot 10^{-35}$ cm^2) at a luminosity of 10^{34} cm^{-2}s^{-1}, the LHC produces two Higgs bosons each second. The number of events observed in each reaction per unit time is:

$$\frac{dN}{dt} = \mathcal{L} \cdot \sigma \cdot BR \tag{14.61}$$

assuming 100% detection efficiency. In this formula, $\mathcal{L}$ is the machine lumi-

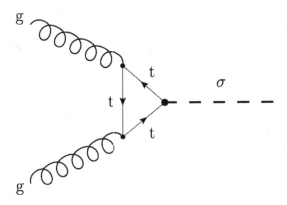

Figure 14.6 Higgs production in proton–proton collisions by the mechanism of inclusive gluon–gluon fusion.

nosity, σ is the Higgs boson production cross section (Figure 14.8) including the eventual kinematic cuts on jets in the final state and BR is the branching ratio (Figure 14.1) into the selected channel.

14.3 THE 125 GEV PARTICLE

The data collected on electroweak particle properties at LEP and other contemporary machines have provided further information on the mass of the Higgs boson, in addition to the lower limit already mentioned.

The Higgs boson mass plays a role in higher order corrections to the mass and decay mode properties of the intermediate bosons, and it is possible to make a fit of the deviations of these properties from the lowest order perturbation theory predictions which depend on the three fundamental constants α, G_F and M_Z (section 7.2) and on the masses of the virtual particles exchanged in the loops. The constraint obtained is still more restrictive following the ob-

Table 14.2 Cross sections (in pb) for the processes of Figure 14.7. $\sigma_{g\text{corr}}$ corresponds to the sum of processes (a)–(c), and σ_{VBFcorr} to process (d). The interference between gluon diagrams and VV fusion is negligible. In the first row the cross sections with minimal cuts are given; in the second row the results with the transverse momentum cuts applied by CMS (the results from ATLAS are similar).

		$\sigma_{g\text{corr}}$	σ_{VBFcorr}	σ_{TOTcorr}
σ, $M_\sigma = 125$ GeV.	Minimal cuts	2.48	1.02	3.50
	CMS cuts	0.149	0.444	0.593

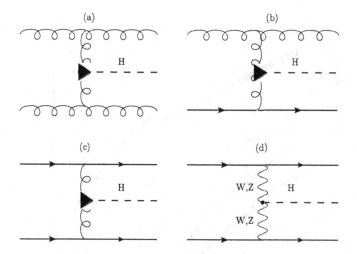

Figure 14.7 Higgs boson production with two jets; gg fusion initiated by gg, gq and qq states, (a)–(c), and fusion of two vector bosons (d).

servation of the t quark, which has removed the penultimate unknown mass, leaving the Higgs boson mass as the only unknown parameter.

Figure 14.9 shows the value of $\Delta\chi^2$, a quantity whose difference from a zero value is an indication of the reliability of the fit, as a function of the mass of the Higgs boson. It is noteworthy that the best fit was obtained for a value of the mass which, at that time, had already been excluded by the LEP measurements, whose limits are indicated by the vertical yellow band which ends at the value $M_\sigma = 114$ GeV. More precisely:

- the most probable value is $M_\sigma = 77^{+69}_{-39}$ GeV;

- $M_\sigma < 188$ GeV, at 95% confidence level.

With this background, the first data taking period at the LHC with significant luminosity started in summer 2011. The search concentrated on the two processes which lead to a peak in the mass distribution of the observed particles, the four leptons in the reaction:

$$p + p \to \sigma + \cdots \to Z + Z^* + \cdots \to l^+ + l^- + l'^+ + l'^- + \cdots \qquad (14.62)$$

and the two photons in the reaction:

$$p + p \to \sigma + 2 \text{ jet} + \cdots \to \gamma + \gamma + 2 \text{ jet} + \cdots \qquad (14.63)$$

One year later the ATLAS [64] and CMS [65] collaborations announced the observation of a new particle at a mass of about 125 GeV which decays

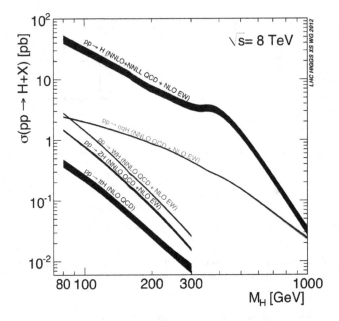

Figure 14.8 QCD predictions for the inclusive production cross sections of the Higgs boson at the LHC vs. its mass, M_H. Source: LHC Higgs cross section Working Group [63].

via the channels (14.62) and (14.63), with weaker evidence also in WW^*, a less prominent signal because of the neutrinos present in the W decays[2].

Figures 14.10–14.11 and 14.12–14.13 show the peaks observed by the collaborations into the channels (14.63) and (14.62), overall a signal of 5 standard deviations for each collaboration. Figures 14.14 and 14.15 make a comparison of the signals observed in different channels to the predictions of the Standard Theory, equation (14.61).

More precise data are required to firmly identify the 125 GeV particle as the Higgs boson, in particular by the determination of its spin and branching ratios into different channels. However, its observation constitutes an extraordinary success for the LHC, for the collaborations which constructed the sophisticated detector systems and for CERN, which has made all this possible.

The coming years will tell us if the intuition of a common origin of the electromagnetic and weak interactions, a thread which links almost one hundred years of research from the work of Enrico Fermi onward, can be considered a definitive picture of the modern physics of particles.

---

[2]At the same time, the collaborations which work at the Fermilab Tevatron [66] announced evidence, in the same energy region, for a deviation of almost 3σ from the Standard Theory prediction in the $b\bar{b}$ channel.

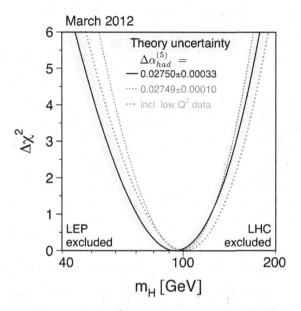

Figure 14.9 The variation of the χ^2 parameter with respect to its value at the minimum, obtained from the Standard Theory fit to the electroweak observables as function of the Higgs boson mass. The shaded areas correspond to the regions excluded by the LEP and by the LHC experiments prior to March 2012. LEP ElectroWeak Working Group, 2012.

Note added

The data on the 125 GeV particle discussed in this book are essentially those presented at the discovery announcement on 4 July 2012. Since then, the LHC ran at a centre of mass energy of 7 to 8 TeV until 13 February 2013, collecting about three times as much data. After intensive refurbishing, LHC recently resumed operation in June 2015 at a centre of mass energy of 13 GeV and improved luminosity which should reach about $2 \cdot 10^{34}$ cm^{-2}sec^{-1}.

Analysis of data obtained in the 2010-2013 runs has continued up to the present (July 2015) and has led to an important consolidation of the picture, confirming the 125 GeV particle as a very serious candidate to fill the role of the Higgs–Brout–Englert boson. In particular:

- Angular correlations of the decay products provide evidence for this particle to have spin-parity $J^P = 0^+$, as required, and exclude alternative assignments allowed by the $\gamma\gamma$ decay such as $J^P = 2^+$.

- The branching ratios of observed decay channels continue to agree with the Standard Theory prediction, with somewhat reduced errors with respect to the figures reported here, although the precision has not improved dramatically.

- Evidence for the important decay modes into $\tau^+\tau^-$ and $b\bar{b}$ is emerging.

The search for new particles, which would indicate the need of an extension of the Standard Theory, has continued, but no positive signal has been reported.

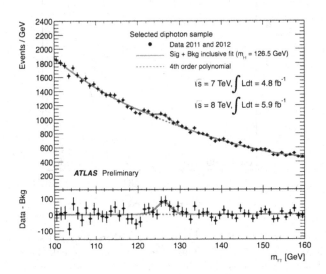

Figure 14.10   ATLAS collaboration, mass distribution of two photons in the reaction (14.63). (F. Gianotti, CERN seminar 4 July 2012); see [64].

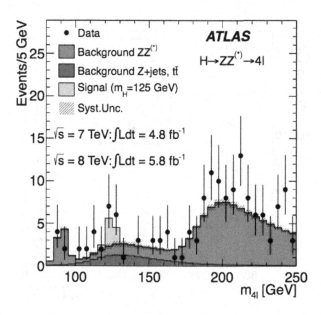

Figure 14.11 ATLAS collaboration, four-lepton mass distribution in the reaction (14.62); see [64]. The new particle is associated with the elevated points of the histogram shaded in lighter gray.

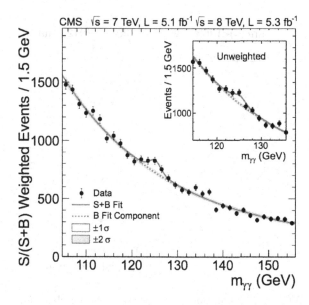

Figure 14.12 CMS collaboration, mass distribution of two photons in the reaction (14.63); see [65].

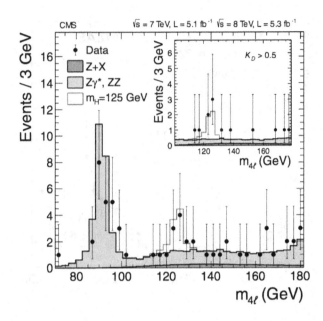

Figure 14.13 CMS collaboration, four-lepton mass distribution in the reaction (14.62); see [65]. The shaded histogram is the prediction of the Standard Theory without the Higgs boson. The unshaded histograms shows the effect of a Higgs boson with 125 GeV mass.

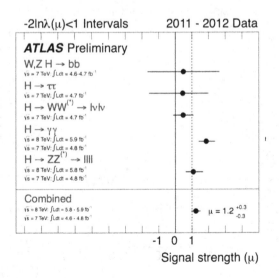

Figure 14.14 ATLAS collaboration, ratio between the number of events observed and predicted by the Standard Theory. (F. Gianotti, CERN seminar 4 July 2012); see [64].

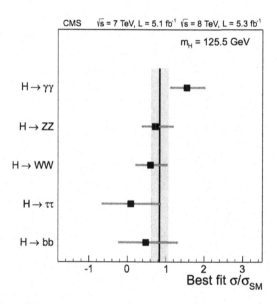

Figure 14.15 CMS collaboration, ratio between the number of events observed and predicted by the Standard Theory; see [65].

Representation of the Two-Point Function

CONTENTS

A.1   CREATION AND DESTRUCTION OPERATORS

It is useful to formulate the theory of the scalar free field in terms of creation and destruction operators which satisfy a normalisation condition adapted to the limit of an infinite volume, in alternative to quantization in a finite box [1].

We start from the fields in a cube of volume V with periodic boundary conditions. We define a set of solutions of the Klein–Gordon (K–G) equation for positive frequency:

$$f_q(x) = \frac{1}{\sqrt{2\omega(\mathbf{q})V}}e^{-iqx}; \quad (\Box + m^2)f_q = 0;$$

$$\mathbf{q} = \frac{2\pi}{L}(n_1, n_2, n_3); \quad \omega(\mathbf{q}) = +\sqrt{\mathbf{q}^2 + m^2}. \tag{A.1}$$

where $n_{1,2,3}$ are integers and L is the side of the cube; $V = L^3$. The functions (A.1) are normalised in V as:

$$\int_V d^3x \; f_q^*(x) i \overleftrightarrow{\partial}_t f_{q'}(x) = \delta_{q,q'} \tag{A.2}$$

with the abbreviation:

$$f \overleftrightarrow{\partial}_t g = f(\partial_t g) - (\partial_t f)g. \tag{A.3}$$

We can construct the creation and destruction operators starting from ϕ

and from f_q as:

$$a_q = \int d^3x \; \left[f_q(x)^* i \overset{\leftrightarrow}{\partial}_t \phi(x) \right]; \quad \text{(destruction)}$$

$$a_q^\dagger = (a_q)^\dagger; \quad\quad\quad\quad\quad\quad \text{(creation)}$$

$$\left[a_q, a_{q'}^\dagger \right] = i\delta_{q,q'}. \quad\quad\quad\quad\quad (A.4)$$

After the spatial integration, the operators a_q could depend on time, but it is straightforward to show that in fact they are constant, as a result of the K–G equation obeyed by ϕ and f_q:

$$\partial_t a_q = i \int d^3x \; \left[f_q(x)^*(\partial_t^2 \phi(x)) - (\partial_t^2 f_q(x)^*)\phi(x) \right] =$$

$$= i \int d^3x \; \left[f_q(x)^*(\Box \phi(x)) - (\Box f_q(x)^*)\phi(x) \right] =$$

$$= i \int d^3x \; \left[f_q(x)^*(\Box + m^2)\phi(x) \right] = 0 \quad\quad (A.5)$$

To switch to the continuum limit, we introduce the projection operator of the momentum states within a three-dimensional interval, $\Delta^3 n$:

$$P = \sum_{\Delta^3 n} |p\rangle\langle p|. \quad\quad\quad\quad\quad (A.6)$$

$P^2 = P$, as a result of the orthonormality of the states $|p\rangle$. If now we pass to the limit of infinite volume, we obtain:

$$P = \int_{\Delta^3 p} |p\rangle \frac{V \, d^3p}{(2\pi)^3} \langle p|. \quad\quad\quad\quad (A.7)$$

Equation (A.7) suggests the definition of kets normalised in the continuum:

$$|\tilde{p}\rangle = \sqrt{\frac{V}{(2\pi)^3}} |p\rangle \quad\quad\quad\quad (A.8)$$

for which:

$$P = \int_{\Delta^3 p} |\tilde{p}\rangle \, d^3p \, \langle\tilde{p}|. \quad\quad\quad\quad (A.9)$$

The condition $P^2 = P$ requires the normalisation condition of the new kets:

$$\langle\tilde{p'}|\tilde{p}\rangle = \delta^{(3)}(p' - p). \quad\quad\quad\quad (A.10)$$

Destruction and creation operators which correspond to the new states are, clearly:

$$\tilde{a}_p = \sqrt{\frac{V}{(2\pi)^3}} a_p; \quad \tilde{a}_p^\dagger = (\tilde{a}_p)^\dagger; \quad\quad\quad (A.11)$$

and the new commutation rules are obtained from (A.10):

$$\delta^{(3)}(p' - p) = \langle 0|\tilde{a}_{p'}\tilde{a}_p^\dagger|0\rangle = \langle 0|\left[\tilde{a}_{p'}, \tilde{a}_p^\dagger\right]|0\rangle; \tag{A.12}$$

or:

$$\left[\tilde{a}_{p'}, \tilde{a}_p^\dagger\right] = \delta^{(3)}(p' - p). \tag{A.13}$$

The field expansion is now written as:

$$\phi(x) = \sum_p \frac{1}{\sqrt{2\omega(\mathbf{p})V}}\left[a_p\, e^{-ipx} + a_p^\dagger\, e^{ipx}\right] =$$

$$= \int \frac{d^3p}{(2\pi)^{3/2}} \frac{1}{\sqrt{2\omega(\mathbf{p})}}\left[\tilde{a}_p\, e^{-ipx} + \tilde{a}_p^\dagger\, e^{ipx}\right]. \tag{A.14}$$

The new states are normalised to have, rather than one particle in the reference volume, a constant particle density. This is seen by calculating the energy of the field, which is equal to (we introduce here the normal product of the operators, as defined in [1]):

$$H = \int d^3x\, \frac{1}{2}\left[: (\partial_t\phi)^2 : + : (\nabla\phi)^2 : +m^2 : \phi^2 :\right] =$$

$$= \int d^3p\, \omega(\mathbf{p})\, \tilde{a}_p^\dagger\tilde{a}_p;$$

$$E_p = \langle\tilde{p}|H|\tilde{p}\rangle = \delta^{(3)}(0)\omega(\mathbf{p}) = \frac{V}{(2\pi)^3}\omega(\mathbf{p}). \tag{A.15}$$

It can therefore be seen that the particle density in the state $|\tilde{p}\rangle$ is $\rho = 1/(2\pi)^3$.

Finally, we note that the relation which links the fields to the destruction operators is written:

$$\tilde{a}_q = \int d^3x\, \left[\tilde{f}_q(x)^* i\overset{\leftrightarrow}{\partial}_t\phi(x)\right];$$

$$\tilde{f}_q(x) = \frac{1}{(2\pi)^{3/2}\sqrt{2\omega(\mathbf{p})}}e^{-ipx}. \tag{A.16}$$

In what follows, we will adopt the continuum normalisation, for brevity omitting the tilde on the operators and on the functions f_q.

A.2 KALLEN–LEHMAN REPRESENTATION

In a field theory with interactions it is not possible to calculate the Green's functions exactly, but the requirement of invariance under Lorentz transformations and a reasonable hypothesis for the structure of the states of one or more particles allows to establish a very useful *spectral representation* of the two-point function.

We write, for $x^0 > 0$:

$$\langle 0|T\,(\phi(x)\,\phi(0))\,|0\rangle = \langle 0|\phi(x)\,\phi(0)|0\rangle = \sum_\alpha \langle 0|\phi(x)|\alpha\rangle\langle\alpha|\,\phi(0)|0\rangle. \quad (A.17)$$

One particle states $|\mathbf{p}\rangle$ and states with two or more particles can contribute to the sum over intermediate states, therefore we divide the sum (and the result) into two parts:

$$\langle 0|\phi(x)\,\phi(0)|0\rangle = \langle 0|\phi(x)\,\phi(0)|0\rangle_1 + \langle 0|\phi(x)\,\phi(0)|0\rangle_{(>1)}. \quad (A.18)$$

The contribution of the one-particle states can be written explicitly, cf. equation (A.9):

$$\langle 0|\phi(x)\,\phi(0)|0\rangle_1 = \int d^3p \langle 0|\phi(x)|\mathbf{p}\rangle\langle\mathbf{p}|\phi(0)|0\rangle. \quad (A.19)$$

Now we proceed in two steps:

- The x dependence is obtained from the relation:

$$\phi(x) = e^{iPx}\,\phi(0)\,e^{-iPx}; \quad (A.20)$$

where P_μ are the operators which represent the total four-momentum, from which:

$$\langle 0|\phi(x)|p\rangle = e^{-ipx}\,\langle 0|\phi(0)|\mathbf{p}\rangle. \quad (A.21)$$

- The matrix element of $\phi(0)$ is parameterised as:

$$\langle 0|\phi(0)|\mathbf{p}\rangle = \frac{\sqrt{Z(p)}}{\sqrt{(2\pi)^3\,2\omega(\mathbf{p})}}; \quad (A.22)$$

for which:

$$\langle 0|\phi(x)\,\phi(y)|0\rangle_1 = \int d^3p \frac{Z(p)}{(2\pi)^3\,2\omega(\mathbf{p})}e^{-ipx}. \quad (A.23)$$

The crucial point is that $Z(p)$ defined by (A.22) *is Lorentz invariant*. It must therefore be a function of the single invariant which we can construct with the four-momentum of the particle, i.e. a function of $p_\mu p^\mu = m^2$ which is a constant, independent of the value of $\mathbf{p}$. The correctness of this statement can be justified, in an intuitive but substantially correct way, as follows:

1. The Green's function is Lorentz invariant and hence so is its restriction to one-particle intermediate states, the left hand side of (A.23).

2. In the right hand side of (A.23), the differential volume $d^3p/(2\omega(\mathbf{p}))$ is Lorentz invariant and so is the exponential e^{-ipx}, from which, to obtain an invariant result, the invariance of $Z(p)$ follows.

We can therefore take $Z(p) = Z(m^2) = Z$ outside the integral and obtain:

$$\langle 0|\phi(x)\,\phi(0)|0\rangle_1 = Z \int d^3p \frac{1}{(2\pi)^3\,2\omega(\mathbf{p})} e^{-ipx}; \quad (x^0 > 0). \tag{A.24}$$

If we repeat these steps in the case $x^0 < 0$, we obtain for the contribution of the one-particle states:

$$\langle 0|\,T\,(\phi(x)\,\phi(0))\,|0\rangle_1 =$$

$$\frac{Z}{(2\pi)^3} \int d^3p \frac{e^{i\vec{p}(\vec{x})}}{2\omega(\mathbf{p})} \left(e^{-i\omega(\mathbf{p})x^0}\theta(x^0) + e^{i\omega(\mathbf{p})x^0}\theta(-x^0) \right)$$

and, comparing with the definition of the Feynman propagator given in [1], we obtain:

$$\langle 0|\,T\,(\phi(x)\,\phi(0))\,|0\rangle_1 = iZ\,D_F(x;m) \tag{A.25}$$

where we have introduced the notation $D_F(x;m)$ to denote the Feynman propagator for a particle of mass m. The constant Z is known as the *renormalisation constant* of the field.

The states of two or more particles can be characterised by the total momentum $\mathbf{p}$, the invariant mass M, and the total energy $E = \sqrt{M^2 + \mathbf{p}^2}$. Unlike the states of a single particle, which correspond to a precise value of m, the states of two or more particles display a continuous spectrum of values of M starting from a certain threshold M_s. For example, the two-particle states of zero total momentum, for which $E = M$, will be composed of two particles of opposite momentum, $\pm\mathbf{q}$, and therefore $M = E = 2\sqrt{m^2 + \mathbf{p}^2} \geq M_s = 2m^1$. The states which contribute to the sum (A.17) are created by ϕ which operates on the vacuum, and they have zero intrinsic angular momentum. Therefore the considerations concerning the states of a single particle apply also to these states, and the contribution of the states of mass M are proportional to $iD_F(x;M)$.

We can therefore give the general expression for the two-point function in a scalar theory:

$$\langle 0|\,T\,(\phi(x)\,\phi(0))\,|0\rangle = iZD_F(x;m) + i\int_{M^2=M_s^2}^{\infty} dM^2\sigma(M^2)D_F(x;M) \tag{A.26}$$

which depends on only two unknown quantities: the normalisation constant Z and the function $\sigma(M^2)$ which is known as the *spectral function*.

We note that in perturbation theory we can expand the two-point function in powers of the coupling constant (λ, in the scalar theory we have chosen as a model in this chapter). Therefore both the renormalisation constant Z and the spectral function $\sigma(M^2)$ should be understood as a series of powers of the coupling constant. In zeroth order this reduces to the free theory, i.e.

$$Z = 1; \quad \sigma(M^2) = 0 \quad \text{(zeroth order of perturbation theory)}. \tag{A.27}$$

---

[1] In the $\lambda\phi^4$ theory, Green's functions with odd numbers of points are not possible and transitions between states with an even number and an odd number of particles are forbidden; the effective threshold is $M_s = 3m$.

More formally ... We can give a formal construction of the second term in (A.26) in the following way.

We consider first the case $x^0 > 0$. The contributions of the states with more than one particle are written:

$$\langle 0|\phi(x)\phi(0)|0\rangle_{>1} = \sum_{n>1}\langle 0|\phi(x)|n\rangle\langle n|\phi(0)|0\rangle =$$

$$= \sum_{n>1} e^{-iP_n x}\, \langle 0|\phi(0)|n\rangle\langle n|\phi(0)|0\rangle. \qquad (A.28)$$

The states $|n\rangle$ are states of two or more particles, characterised by the total four-momentum P_n, plus other quantum numbers which we do not need to specify. We insert into (A.28) two δ-functions, which integrate explicitly to unity:

$$\sum_n \langle 0|\phi(x)|n\rangle\langle n|\phi(0)|0\rangle =$$

$$= \int dM^2 \int \frac{d^4p}{(2\pi)^4}\, \delta(p^2 - M^2) \times$$

$$\sum_n (2\pi)^4 \delta^{(4)}(p - P_n)e^{-iP_n x}\, \langle 0|\phi(0)|n\rangle\langle n|\phi(0)|0\rangle =$$

$$= \int dM^2 \int \frac{d^4p}{(2\pi)^4}\, \delta(p^2 - M^2)e^{-ipx} \times$$

$$\sum_n (2\pi)^4 \delta^{(4)}(p - P_n)\, \langle 0|\phi(0)|n\rangle\langle n|\phi(0)|0\rangle. \qquad (A.29)$$

where we have taken the exponential outside the sum using the δ-function.

The crucial point is that, as in the case of a single particle, the sum over states gives a function of four-momentum p which is *Lorentz invariant*. Therefore the sum must be a function of p^2 or, as a result of the second δ-function, of M^2. We can denote this function as:

$$(2\pi)\sigma(M^2) = \sum_n (2\pi)^4 \delta^{(4)}(p - P_n)\, \langle 0|\phi(0)|n\rangle\langle n|\phi(0)|0\rangle \quad (M^2 = p^2) \text{ (A.30)}$$

and obtain:

$$\langle 0|\phi(x)\phi(0)|0\rangle_{>1} = \int dM^2 \sigma(M^2) \int \frac{d^4p}{(2\pi)^4}\, e^{-ipx} =$$

$$= \int dM^2 \sigma(M^2)\frac{1}{(2\pi)^3} \int \frac{d^3p}{2\omega(p, M)}\, e^{-i(\omega(p,M)t}e^{i\mathbf{p}\cdot\mathbf{x}} \quad (A.31)$$

where $\omega(p, M)$ is the energy which corresponds to a particle of four-momentum p and mass M.

We repeat the argument for $x^0 < 0$ and change the integration variable $\mathbf{p} \to -\mathbf{p}$. We find:

$$\langle 0|\phi(0)\phi(x)|0\rangle_{>1} = \int dM^2\sigma(M^2) \int \frac{d^4p}{(2\pi)^4} \, e^{+ipx} =$$

$$= \int dM^2\sigma(M^2) \frac{1}{(2\pi)^3} \int \frac{d^3p}{2\omega(p,M)} \, e^{+i\omega(p,M)t} e^{i\mathbf{p}\cdot\mathbf{x}}. \quad (A.32)$$

In conclusion:

$$\langle 0|T\left[\phi(x)\phi(0)\right]|0\rangle_{>1} = \int dM^2\sigma(M^2) \times$$

$$\left\{ \frac{1}{(2\pi)^3} \int \frac{d^3p}{2\omega(p,M)} \, e^{i\mathbf{p}\cdot\mathbf{x}} \left[\theta(x^0)e^{-i\omega(p,M)t} + \theta(-x^0)e^{+i\omega(p,M)t}\right] \right\}. \quad (A.33)$$

Comparing with equation (A.25) we finally find:

$$\langle 0|T\left[\phi(x)\phi(0)\right]|0\rangle_{>1} = i\int dM^2\sigma(M^2)D_F(x,M). \quad (A.34)$$

The function $\sigma(M^2)$ is different from zero only for those values of M^2 which correspond to the mass of some possible intermediate state, i.e. for $M^2 > (2m)^2$, therefore the integration limits go from $(2m)^2$ to ∞ and we find again the second term of (A.26)[2].

---

[2] As already noted, in the $\lambda\phi^4$ theory, the intermediate state can contain 1, 3, 5, in general $2k + 1$, particles; therefore the minimum mass is $3m$.

Decomposition of Finite-Dimensional Matrices

CONTENTS

B.1 THEOREM

An arbitrary complex matrix H can be written in the form:

$$H = MV \tag{B.1}$$

where M is a Hermitian matrix with positive or zero eigenvalues and V is a unitary matrix[1].

The form (B.1) generalises to matrices the well known polar factorisation of a complex number z, according to which:

$$z = \rho e^{i\phi}, \quad \rho \geq 0. \tag{B.2}$$

We recall that the factorisation (B.2) is obtained directly by writing:

$$z^2 = (zz^*)(\frac{z}{z^*}) = \rho^2 e^{2i\phi} \tag{B.3}$$

and extracting the square root.

B.2 PROOF

The formula (B.3) cannot be extended directly to matrices, since in general H and $H^\dagger$ do not commute and therefore, even assuming that H is non-singular, the expression $(H^\dagger)^{-1}H$ does not provide a unitary matrix, precisely

---

[1]The original derivation is due to Cabibbo and Gatto and by Feinberg, Kabir and Weinberg [67].

because of the non-commutativity of the two matrices. Restricting ourselves to the case of a non-singular matrix H, we can however demonstrate that, even though $HH^\dagger$ and $H^\dagger H$ may not commute, these two matrices obey the same characteristic equation and therefore have the same spectrum of eigenvalues.

Let us begin with the case in which H is a non-singular matrix. Defining:

$$M_a^2 = HH^\dagger; \quad M_b^2 = H^\dagger H; \tag{B.4}$$

we find:

$$det\left(M_a^2 - \lambda\right) = det\left(H^{-1}H\right) \times det\left(M_a^2 - \lambda\right) =$$
$$= det\left[H^{-1}\left(HH^\dagger - \lambda\right)H\right] = det\left(M_b^2 - \lambda\right).$$

Therefore the matrices M_a^2 and M_b^2 can be diagonalised into the same matrix by two, generally different, unitary transformations:

$$M_a^2 = U'\rho^2(U')^\dagger; \quad M_b^2 = V'\rho^2(V')^\dagger \tag{B.5}$$

where ρ^2 is diagonal with positive elements. The general case is obtained by continuity and ρ can also have some vanishing diagonal values.

We now construct:

$$h = (U')^\dagger HV'. \tag{B.6}$$

Clearly, we have:

$$hh^\dagger = (U')^\dagger HH^\dagger U' = (U')^\dagger M_a^2 U' = \rho^2;$$
$$h^\dagger h = (V')^\dagger H^\dagger HV' = (V')^\dagger M_b^2 V' = \rho^2;$$

therefore h and $h^\dagger$ commute with each other:

$$hh^\dagger = h^\dagger h. \tag{B.7}$$

We can then apply the same reasoning which leads to (B.2), by writing:

$$h^2 = \left(hh^\dagger\right)\left(\frac{h}{h^\dagger}\right) \tag{B.8}$$

and it is easily shown that the second set of parentheses defines a unitary matrix:

$$\left[h(h^\dagger)^{-1}\right]^\dagger = \left(h^{-1}h^\dagger\right) = \left(h^\dagger h^{-1}\right) = \left[h(h^\dagger)^{-1}\right]^{-1}. \tag{B.9}$$

We can now extract the square root of (B.8), an allowed operation for positive Hermitian matrices and unitary matrices, to obtain:

$$h = \rho W \tag{B.10}$$

with ρ positive and W unitary. Finally, comparing with equation (B.6), we obtain:

$$H = U'h(V')^\dagger = U'\rho W(V')^\dagger = U'\rho(U')^\dagger U'W(V')^\dagger = MV \tag{B.11}$$

as required.

Corollary. Any complex matrix H can be diagonalised with a transformation which involves two unitary matrices (bi-unitary transformation) of the form:

$$H = U\rho V^{\dagger} \tag{B.12}$$

where ρ is a diagonal matrix with positive or zero elements, and U and V are two unitary matrices.

The representation (B.12) has been widely employed in the literature on weak interactions, starting from the 1960s.

Continuous Groups and Their Representations

CONTENTS

C.1   DEFINITION OF A GROUP

A *group* is a set of elements G conforming to a multiplication law which satisfies the following requirements:

- the law of multiplication is associative: $s(ht) = (sh)t$,

- an identity element exists for multiplication from the left: $se = s$, for all s,

- every element s has an inverse from the left which we denote as s^{-1}, such that:

$$s^{-1}s = e. \tag{C.1}$$

Note 1.   From the preceding definitions it follows that:

$$es = s^{-1}(se)s = s^{-1}(s)s = s; \tag{C.2}$$

therefore the identity elements for multiplication from left and right are the same:

$$es = s \text{ for all } s. \tag{C.3}$$

Similarly, the inverse elements from left and right are also the same.

Note 2. We have placed no requirement on the commutation of the product and, in general, $sh \neq hs$. The groups for which $sh = hs$ for all elements are called *commutative* or *abelian* groups.

Note 3. A natural group of transformations of G in G exists, defined by the conjugation operation with a fixed element f:

$$g' = T_f(g) = fgf^{-1}. \tag{C.4}$$

Note 4. In what follows we will limit ourselves to groups which are continuous, differentiable and compact. Why? The first two conditions are clear and natural: we are interested in symmetries under continuous transformations on the field variables, the extension of rotations and internal transformations like isotopic spin. In quantum mechanics, we are furthermore interested in representations of these symmetries by unitary operators, which preserve the metric in the Hilbert space of quantum states. Now, as we will see, the irreducible representations of a compact group are unitary and finite dimensional, within equivalences. Therefore symmetries based on compact groups give rise to finite-dimensional particle multiplets which appear in nature as groups of particles with similar masses and equal spatial quantum numbers like spin, parity and charge conjugation.

C.2   LIE ALGEBRAS

The elements of continuous groups can be parameterised one to one with a certain number of variables $\alpha_1, \alpha_2, \ldots, \alpha_n$ (in certain cases more mappings are necessary to parameterise the whole range). In particular we can parameterise an infinitesimal region around e in such a way that within it the parameters $\alpha_1, \alpha_2, \ldots, \alpha_n$ are infinitesimal, with $g(0, 0, \ldots, 0) = e$.

This region can be considered a vector space (the *tangent space* to the manifold) and on it the structure of the transformations (C.4) is reflected in the structure of an algebra of commutators, the generators of the infinitesimal transformations:

$$f \to 1 + i \sum_i \alpha_i T^i; \qquad g \to 1 + i \sum_i \beta_i T^i;$$

$$g \to g' = fgf^{-1} \to 1 + i \sum_i \alpha_i T^i - (\alpha_i \beta_j)[T^i, T^j];$$

$$[T^i, T^j] = T^i T^j - T^j T^i. \tag{C.5}$$

Because g' also belongs to the tangent space, the commutator must return a combination of generators, for which:

$$[T^i, T^j] = if^{ijk}T^k \tag{C.6}$$

(with summation over repeated indices). The constants f^{ijk} (structure constants) are characteristic of the algebra and we can choose the generators in such a way that the f are completely antisymmetric in their three indices. For continuous groups, the structure of the infinitesimal transformations (C.4) is that of a *Lie algebra*. The *rank* of a Lie algebra is the maximum number of generators which commute with each other and that therefore can be simultaneously diagonalised. Clearly the rank is ≥ 1.

As well as being antisymmetric, the structure constants in (C.6) should satisfy algebraic relations which follow from the Jacobi identity, valid for commutators of arbitrary matrices:

$$[X, [Y, Z]] + [Y, [Z, X] + [Z, [X, Y]] = 0. \tag{C.7}$$

Equation (C.7) can be proved by explicitly writing the commutators. Substituting (C.6) into (C.7), we find:

$$\sum_s (f^{ijs}\, f^{ksm} + f^{jks}\, f^{ism} + f^{kis}\, f^{jsm}) = 0. \tag{C.8}$$

Analysis of the possible realisations of the structure constants which satisfy equations (C.6) and (C.7) leads to the Cartan classification of the simple Lie algebras.

Note. In the case of the rotation group, the structure of the Lie algebra is provided by the well known commutation rules:

$$[J^i, J^j] = i\, \epsilon^{ijk} J^k \quad (i, j, k = 1, 2, 3). \tag{C.9}$$

ϵ^{ijk} is the completely antisymmetric tensor in three dimensions. The three angular momentum generators do not commute with each other, so the algebra characterised by (C.9) has rank 1.

Universal covering group. Any element of the group can be reached from the infinitesimal generators; the algebra therefore determines the group, unless there are problems which can arise if the group G is not simply connected. Actually, each element of the group can be approximated by the product of elements which are all in a neighbourhood around e. The problem is that this expression is not unique. In general, we can "deform" the chain continuously and therefore show that the result does not depend on the chain chosen. However, this is not true if, while changing g in the group, we turn around a singularity, as indeed happens if the group is not simply connected. Given a

Lie algebra , L, however, it is always possible to find a simply connected group, $\bar{G}$ (*universal covering group*) for which L is the Lie algebra of the infinitesimal generators. There could be another non-simply connected group, G, which has the same infinitesimal algebra. In this case a homomorphism of $\bar{G}$ in G exists which reduces to a true and exact isomorphism in a sufficiently small neighbourhood around the identity. The representations for one value of $\bar{G}$ can be representations for many values of G. This is the case for $SU(2)$, (matrices U, 2x2, unitary, det(U)=1) which is the covering group of the algebra (C.9) and of $O(3)$, the group of rotations in a three-dimensional Euclidean space. $O(3)$ has the same algebra as $SU(2)$ but is not simply connected.

In quantum mechanics we are interested in unique representations at least to within a phase factor (if a 2π rotation is applied to the ket $|a>$ which leads to $-|a>$ rather than $|a>$, the effect is nevertheless that of returning to the same physical starting state). In the case of rotations, this corresponds to taking the representation of the covering group, $SU(2)$, rather than being restricted only to the representations of $O(3)$ which would exclude semi-integer spins.

C.3   REPRESENTATION OF A GROUP

A representation of the group G is a mapping between the elements of the group and linear operators in a vector space, L, which preserves the multiplication law of G:

$$g \rightarrow T(g);$$
$$gh \rightarrow T(gh) = T(g)T(h); \tag{C.10}$$

from which it follows that $T(e) = 1$. The operators T should not be singular, since, if:

$$g \rightarrow T(g); \quad g^{-1} \rightarrow T(g^{-1}), \tag{C.11}$$

it should also be true that:

$$T(g)T(g^{-1}) = T(gg^{-1}) = T(e) = 1; \tag{C.12}$$

from which:

$$T(g^{-1}) = T^{-1}(g). \tag{C.13}$$

Some definitions:

- Two representations $T_1(g)$ in L_1, $T_2(g)$ in L_2, are said to be *equivalent* if there exists a non-singular operator A which transforms L_1 into L_2 such that:

$$AT_1(g) = T_2(g)A \quad \text{for all } g \text{ in G.} \tag{C.14}$$

Equivalent representations are the same in all respects. Within a given class of equivalence we can choose the one which best suits our purpose.

- A representation is said to be unitary if $T(g)$ are unitary matrices, for which:

$$T(g^{-1}) = T^{-1}(g) = [T(g)]^{*T} = T(g)^{\dagger}. \tag{C.15}$$

($* =$ complex conjugation, T $=$ transposition, $^{\dagger} =$ Hermitian conjugate.)

- *Reducible* representations: the matrices $T(g)$ allow an invariant subspace $V \neq L, V \neq 0$:

$$T(g)x \ \epsilon \ V \quad \text{if } x \ \epsilon \ V \text{ for all g.} \tag{C.16}$$

- Completely reducible representation: the matrices $T(g)$ are *in blocks*. More precisely:

$$L = \oplus_{i,\alpha} L_{i,\alpha};$$
$$T(g) = \oplus_{i,\alpha} T^{(i,\alpha)}(g). \tag{C.17}$$

The symbol $\oplus_{i,\alpha}$ denotes the direct sum over irreducible vector spaces $L_{i,\alpha}$ and the direct sum of matrices, each of which acts on the vectors of $L_{i,\alpha}$. The matrices $T(g)$ are therefore composed of diagonal blocks, $T^{(i,\alpha)}(g)$. The index i characterises the non-equivalent irreducible representations which appear in the reduction of $T(g)$, the index α distinguishes between equivalent reducible representations, and is necessary if there are degeneracies in the reduction of $T(g)$.

We should underline that, in general, a representation can be reducible (have a non-trivial space) but not completely reducible (composed of blocks; see [69], for example, for more details). The necessary and sufficient condition for that to occur is that the space V_{ort}, orthogonal to the invariant space V, should also be invariant.

However:

- Unitary representations are always completely reducible.

Proof. We choose $x \ \epsilon \ V, y \ \epsilon \ V_{ort}$, where V_{ort} is completely orthogonal to V and therefore $(x, y) = 0$. Since V is invariant, for every g:

$$0 = (T(g^{-1})x, y) = (T(g)^{\dagger}x, y) = (x, T(g)y). \tag{C.18}$$

Therefore, if $y \ \epsilon \ V_{ort}$, also $T(g)y \ \epsilon \ V_{ort}$ and the matrices $T(g)$ are block-diagonal in $L = V \oplus V_{ort}$.

Note. In the case of the rotation group, the irreducible representations are those of fixed angular momentum, J (therefore $i = J$), and the index α serves to distinguish possible components of T with the same angular momentum.

C.4   ALGEBRA OF INFINITESIMAL GENERATORS

A representation of the algebra is a mapping:

$$T^i \rightarrow K(T^i) = K^i \tag{C.19}$$

K^i are linear operators on a vector space L, which provide a realisation of the commutation rules of the algebra.

$$[K^i, K^j] = if^{ijk}K^k. \tag{C.20}$$

Via the operators K^i we can obtain a representation of the elements of the group in an infinitesimal neighbourhood around the identity:

$$g \rightarrow 1 + i\alpha_i T^i;$$
$$T(g) = 1 + i\alpha_i K^i. \tag{C.21}$$

It is possible to write a differential equation in the variables α_i whose solution gives the finite elements of the representation $T(g)$ (cf. [69]). In the case of simply connected groups, and therefore in the case of the covering group, the representations of the Lie algebra completely determine the representations of the group.

The converse is obvious; the operators K^i can be obtained by expanding the matrices $T(g(\alpha_1, \alpha_2, \ldots, \alpha_n))$ around the origin. Note that a unitary representation of G gives rise to a representation of the algebra with Hermitian operators (this is the reason for the appearance of the imaginary number, i, in the preceding equations).

The regular representation.   We can consider the infinitesimal generators as the basis of the vector space generated by their linear combinations:

$$x = i\zeta_i T^i. \tag{C.22}$$

The commutation operation acts as a linear transformation on this space and itself provides a representation of the underlying Lie algebra:

$$x \rightarrow x' = F(T^j)x = [T^j, i\zeta_i T^i] = i\zeta_i(if^{jik}T^k) = i\zeta'_k T^k;$$
$$(\zeta')^k = (K^j_{\text{reg}})_{ki}\zeta^i;$$
$$(K^j_{\text{reg}})_{ki} = -if^{kji}. \tag{C.23}$$

Clearly, we must verify that the matrices K^j_{reg} obey the correct commutation rules, equation (C.20). We leave to the reader the task of confirming that these relations, written explicitly, simply reduce to the Jacobi identity, equation (C.7).

The transformations $x \rightarrow x'$ given above constitute a representation of the algebra which is known as the *regular* or *adjoint representation*. The dimension of the regular representation is clearly the same as the dimension of the Lie algebra.

C.5   REPRESENTATIONS OF COMPACT OR FINITE GROUPS

The situation is explained completely in the classic work of Peter and Weyl [68] (see [69]) and can be summarised in the following way. Assuming that $T(g)$ is continuous:

$$||T(g')x - T(g)x|| \rightarrow 0 \quad \text{if} \quad g' \rightarrow g, \quad \text{for all } x \text{ of } L. \tag{C.24}$$

Peter and Weyl show that:

- In every class of equivalent representations there exists a unitary representation (UR).

- Every irreducible representation is of finite dimensions.

- Every unitary representation is completely reducible.

In view of these results, we can restrict ourselves to the study of finite-dimensional unitary representations. Furthermore, given the results obtained in the previous section, we can restrict ourselves to finite-dimensional representations of the Lie algebra corresponding to the group, with infinitesimal generators K^i which are Hermitian.

C.6   THE EXAMPLE OF $SU(2)$

Representations of the angular momentum algebra (cf. for example [4]) are each characterised by a quantum number J (the total angular momentum) which can take either integer or half-integer values $J = 0, \frac{1}{2}, 1, \frac{3}{2}, 2, \ldots$. The dimension of the representation is $\text{Dim}(J) = 2J + 1$ and the vectors of a complete basis are distinguished by the eigenvalues of J_3, the diagonal component of the total angular momentum:

$$|J, J_3 >; \qquad J_3 = -J, -J + 1, \ldots, +J; \tag{C.25}$$

(sometimes J_3 is referred to as the magnetic quantum number). The operator $\mathbf{J}^2 = J_1^2 + J_2^2 + J_3^2$ commutes with the three angular momentum operators and:

$$\mathbf{J}^2 |J, J_3 >= J(J + 1)|J, J_3 > . \tag{C.26}$$

It is not difficult to find the matrix elements of the non-diagonal generators (cf. [4] again). For convenience, we give the matrix elements of the raising and lowering operators:

$$J^{(\pm)} = (J_1 \pm iJ_2);$$
$$< J, J_3 + 1|J^{(+)}|J, J_3 >= \sqrt{J(J + 1) - J_3(J_3 + 1)};$$
$$< J, J_3 - 1|J^{(-)}|J, J_3 >= \sqrt{J(J + 1) - J_3(J_3 - 1)}. \tag{C.27}$$

The matrices $K_{1,2,3}^{(J)}$ which represent the elements of the algebra $J_{1,2,3}$ are

Hermitian and the representation of the $SU(2)$ group is unitary. To obtain the corresponding matrices it is necessary to extend the parameterisation to finite elements. We can characterise every rotation in the space of three dimensions with a vector of unit length (n_i, $i = 1, 2, 3$) and a rotation angle $0 \leq \alpha \leq 2\pi$. We consider explicitly the matrices of the spin $\frac{1}{2}$ representation, the fundamental representation of $SU(2)$; all others can be obtained by combining these representations with themselves.

$$\mathcal{D}^{(1/2)}(\alpha, \mathbf{n}) = exp[-i\frac{\alpha}{2}\mathbf{n} \cdot \sigma] = \cos\frac{\alpha}{2} - i\mathbf{n} \cdot \sigma \sin\frac{\alpha}{2};$$
$$\mathbf{n} = (\sin\theta\cos\phi, \ \sin\theta\sin\phi, \ \cos\theta).$$

and we have introduced the three Pauli matrices:

$$\sigma_1 = \begin{pmatrix} 0 & 1 \\ 1 & 0 \end{pmatrix}; \ \ \sigma_2 = \begin{pmatrix} 0 & -i \\ i & 0 \end{pmatrix}; \ \ \sigma_3 = \begin{pmatrix} 1 & 0 \\ 0 & -1 \end{pmatrix}.$$

The matrices $\mathcal{D}^{(1/2)}$ are also elements of $SU(2)$, therefore the parameterisation above characterises the corresponding differential manifold.

C.7   TENSOR PRODUCTS: THE CLEBSCH–GORDAN SERIES

Given two vector spaces, L_1 and L_2, we can define a new space which is the tensor product of the two. In Dirac notation:

$$|v, w >= |v > |w >; \quad v \ \epsilon \ L_1, w \ \epsilon \ L_2. \tag{C.28}$$

Given the representations $T_1(g)$ in L_1, and $T_2(g)$ in L_2, the tensor product is defined:

$$T(g) = T_1(g) \otimes T_2(g);$$
$$T(g)|v, w >= T_1(g)|v > T_2(g)|w > . \tag{C.29}$$

(It confirms that $T(g)$ is a representation of G and is unitary, if $T_{1,2}(g)$ are unitary.)

The algebra of the infinitesimal generators is given by:

$$K^i = (K_1)^i \otimes 1 + 1 \otimes (K_2)^i. \tag{C.30}$$

In general, the representation T is not irreducible. For compact groups, however, the representation can be completely reduced into irreducible diagonal blocks. In analogy with angular momentum, we describe the vectors of an irreducible UR with a set of quantum numbers, j, which characterise the representation, and a set of 'magnetic' quantum numbers, m, which characterise the vector within the representation j. For an algebra of rank R, there will be exactly R magnetic quantum numbers, the eigenvalues of the operators which can be simultaneously diagonalised. We use italic characters for the magnetic numbers to remind us that they are a set of quantum numbers.

In terms of the basis vectors in L_1, L_2, a basis in L is given by the tensor products:

$$|j_1, m_1; j_2, m_2 >= |j_1, m_1 > |j_2, m_2 > . \tag{C.31}$$

On the other hand, after the complete reduction of the representation T, we should be able to write the space L and the matrices T in the form given in (C.17):

$$L = \oplus_{j,\alpha} L_{j,\alpha};$$
$$T(g) = \oplus_{j,\alpha} T^{(j,\alpha)}(g). \tag{C.32}$$

with the same meaning of the symbols (we recall that α distinguishes equivalent representations which, if required, appear in the expansion).

In each space $L_{j,\alpha}$, we choose a base set of vectors $|j, m, \alpha >$. The combination of these bases clearly forms a complete basis in L, an alternative to that given before. A unitary matrix should therefore exist which carries out the change of basis. As a formula:

$$|j_1, m_1 > |j_2, m_2 >= \sum_{j,m,\alpha} C(j_1, m_1; j_2, m_2|j, m, \alpha)|j, m, \alpha >; \tag{C.33}$$

$$|j, m, \alpha >= \sum_{m_1,m_2} C^*(j_1, m_1; j_2, m_2|j, m, \alpha)|j_1, m_1 > |j_2, m_2 > . \tag{C.34}$$

The coefficients C are known as Clebsch–Gordan coefficients and satisfy the orthonormality relations:

$$\sum_{j,m,\alpha} C(j_1, m_1; j_2, m_2|j, m, \alpha)C^*(j_1, m_1'; j_2, m_2'|j, m, \alpha) = \delta_{m_1,m_1'}\delta_{m_2,m_2'}; \tag{C.35}$$

$$\sum_{m_1,m_2} C(j_1, m_1; j_2, m_2|j, m, \alpha)C^*(j_1, m_1; j_2, m_2|j', m', \alpha') = \delta_{j,j'}\delta_{m,m'}\delta_{\alpha,\alpha'}. \tag{C.36}$$

With the help of the previous formulae we can express the matrix elements which represent the group in the form of tensor products in terms of irreducible matrices. We find:

$$(T^{(j_1)}(g))_{m_1,m_1'}(T^{(j_2)}(g))_{m_2,m_2'} =$$
$$= \sum_{j,m,\alpha} C(j_1, m_1; j_2, m_2|j, m, \alpha)(T^{(j)}(g))_{m,m'}C^*(j_1, m_1'; j_2, m_2'|j, m', \alpha). \tag{C.37}$$

Note. In the case of representations of $SU(2)$, the result above reduces to the well known combination of two angular momenta. In this case j, the resultant momentum, is included in the interval:

$$|j_1 - j_2| \geq j \geq |j_1 + j_2|; \tag{C.38}$$

j varies in unit steps, and each j appears only once (the index α is not needed).

The Clebsch–Gordan coefficients for the simplest case (i.e. $\frac{1}{2} \otimes 1$) are tabulated in the Particle Data Group Book [5]. $SU(3)$ Clebsch–Gordan coefficients, for the most commonly used examples, have been published in [71].

As can be seen from the previous formulae, the general case of a compact group does not introduce substantial complications compared to $SU(2)$, except for the fact that the same representation can appear several times in the Clebsch–Gordan series. For example, in the decomposition of the products of three fundamental representations of $SU(3)$, considered in chapter 8, the octet representation appears twice; see equation (8.5).

C.8 SCHUR'S LEMMA

This lemma allows us to characterise the matrix elements of operators which are invariant under group transformations, i.e. operators which commute with the matrices of the representation. We deal first with the case of irreducible representations. We should consider two cases.

1. Given two irreducible and non-equivalent representations, $T_1(g)$ in L_1 and $T_2(g)$ in L_2, let A be an operator which transforms vectors from L_1 into vectors of L_2 and satisfies:

$$AT_1(g) = T_2(g)A, \quad \text{for all } g. \tag{C.39}$$

Lemma I: A is the null operator, $A = 0$.

Proof.   We consider the nucleus of A in L_1, i.e. the set N of vectors from L_1 such that $Ax = 0$. N is clearly an invariant of T_1:

$$AT_1(g)x = T_2(g)Ax = 0, \text{for all } g. \tag{C.40}$$

Since $T_1(g)$ is irreducible, $N = L_1$ or $N = 0$. In the first case the lemma is proven. In the second, we consider in L_2 the image J of the transformation A. J is an invariant of T_2, given that, if $y = Ax$:

$$T_2(g)y = T_2(g)Ax = AT_1(g)x = Ax', \quad \text{that } \epsilon \, J. \tag{C.41}$$

But T_2 is irreducible, therefore $J = 0$ or $J = L_2$. The second case is excluded since T_1 and T_2 are non-equivalent, therefore $J = 0$.
    QED.

2. Given $T_1(g)$ and $T_2(g)$ as irreducible and equivalent representations, let U be a unitary matrix which transforms L_1 into L_2, such that:

$$T_1(g) = U^\dagger T_2(g)U, \quad \text{for all } g. \tag{C.42}$$

Lemma II: Every operator A which satisfies (C.39) is a multiple of U, $A = \lambda U$.

As a preliminary, we note that by substituting equation (C.42) into (C.39), we find:

$$BT_2(g) = T_2(g)B \tag{C.43}$$

with $B = AU^\dagger$.

Clearly, if $A = \lambda U$, it must also be true that $AU^\dagger = B = \lambda 1$ and an equivalent form of the lemma is the following:

Lemma II′: every matrix B which commutes with the matrices $T_2(g)$ of an irreducible representation is a multiple of the identity.

Proof. The matrix B has at least one eigenvector x:

$$Bx = \lambda x, \quad x \neq 0. \tag{C.44}$$

The vector space, V, which belongs to the eigenvalue λ is an invariant space of $T_2(g)$, given that:

$$B[T_2(g)x] = T_2(g)Bx = \lambda[T_2(g)x]. \tag{C.45}$$

Since T_2 is irreducible, $V = 0$ or $V = L_2$. The first case is excluded by the existence of at least one eigenvector. Therefore $V = L_2$ and $B = \lambda 1$.
   QED.

C.9   MATRIX ELEMENTS OF INVARIANT OPERATORS

We consider a quantum system (an atom, or field) whose states are represented by vectors from a Hilbert space, L. An operator $\mathcal{O}$ is invariant under the group transformations if it satisfies the condition:

$$U^\dagger(g)\mathcal{O}U(g) = \mathcal{O}, \tag{C.46}$$

where $U(g)$ are unitary operators which represent the action of the transformations of g on the states of the system. For a compact group, $U(g)$ is completely reducible into finite-dimensional blocks.

$$L = \oplus_{j,\alpha}L_{j,\alpha};$$
$$U(g) = \oplus_{j,\alpha}U^{(j,\alpha)}(g). \tag{C.47}$$

We consider the matrix elements between vectors which belong to two irreducible blocks:

$$< j, \alpha, m|\mathcal{O}|j', \beta, m' > = \mathcal{O}(j\alpha, j'\beta)_{m,m'}. \tag{C.48}$$

From (C.46) we find:

$$
\begin{aligned}
(\mathcal{O}(j\alpha, j'\beta))_{m,m'} &= < j, \alpha, m|\mathcal{O}|j', \beta, m' > = \\
&= < j, \alpha, m|U(g)^\dagger \mathcal{O}U(g)|j', \beta, m' > = \\
&= (T^{(j',\beta)})_{m',n'}(T^{(j,\alpha)})^*_{m,n}(\mathcal{O}(j\alpha, j'\beta))_{n,n'} = \\
&= \left\{[T^{(j,\alpha)}(g)^T]^\dagger\, [\mathcal{O}(j\alpha, j'\beta)]\, [T^{(j',\beta)}(g)^T]\right\}_{m,m'}. \tag{C.49}
\end{aligned}
$$

The finite-dimensional matrix $\mathcal{O}(j\alpha, j'\beta)$ satisfies the conditions of Schur's lemma:

$$\mathcal{O}(j\alpha, j'\beta) \, [T^{(j'\beta)}(g)]^T = [T^{(j,\alpha)}(g)]^T \, \mathcal{O}(j\alpha, j'\beta). \qquad (C.50)$$

From here it follows that:

- $\mathcal{O}(j\alpha, j'\beta) = 0, \quad$ if $j \neq j'$;

- $\mathcal{O}(j\alpha, j'\beta)_{n,n'} = \|\mathcal{O}(j, \alpha, \beta)\| \, \delta_{n,n'}, \quad$ if $j' = j$.

In words, an invariant operator has matrix elements only between equivalent representations; these matrix elements are diagonal in, and independent of, magnetic quantum numbers. There are as many independent coefficients, $\|\mathcal{O}(j, \alpha, \beta)\|$, as there are pairs of equivalent representations.

C.10   INVARIANT MEASURE AND ORTHONORMALITY

For finite groups, the sum over the elements of the group is invariant under the operation which shifts the elements of the group, multiplying them by a fixed element:

$$\sum_g f(g) = \sum_g f(hg) \qquad (C.51)$$

for any function of the elements of the group. It is reasonable to think of introducing a similar measure in the case of a compact group, transporting the measure $d\alpha_1 d\alpha_2 \dots d\alpha_n$ from the neighbourhood of the identity to the neighbourhood of a general element, by means of the multiplication of the group. Indeed (cf. [7]) for compact groups an invariant measure for multiplication from the left can be defined:

$$\int dg f(hg) = \int dg f(g) \qquad (C.52)$$

and the measure remains invariant also for multiplication from the right. The measure (C.52) is uniquely defined, at least to within a multiplicative constant which we choose so that

$$\int dg = 1. \qquad (C.53)$$

We consider two irreducible representations, $T_1(g)$ and $T_2(g)$, and we construct the matrix $B_{q,a}$:

$$B_{q,a} = \int dg \, T_1(g)_{q,r} T_2(g)^*_{a,b} = \int dg \, T_1(g)_{q,r} T_2(g)^\dagger_{b,a}; \qquad (C.54)$$

The matrix $B_{q,a}$ also depends on two indices r and b which for the moment we keep fixed and therefore do not write.

B is invariant, in the sense that:

$$T_1(h)BT_2(h)^\dagger = B; \qquad (C.55)$$

Actually:

$$T_1(h)_{q,q'} \int dg [T_1(g)]_{q',r} [T_2(g)]^\dagger_{b,a'} T_2(h)^\dagger_{a',a} =$$

$$= \int dg [T_1(hg)]_{q,r} [T_2(hg)]^\dagger_{b,a} =$$

$$= \int dg \, T_1(g)_{q,r} T_2(g)^\dagger_{b,a} = B_{q,a}. \qquad (C.56)$$

On the basis of Schur's lemma, we can conclude the following:

• if T_1 and T_2 are non-equivalent, $B = 0$;

• if $T_1 = T_2$, $B_{a,q} = \lambda(r,b,j)\delta_{a,q}$, where j characterises the representation. In the second case, we can put $a = q$ and sum, obtaining:

$$\lambda(r,b,j)\mathrm{Dim}(j) = \int dg (T_2(g))^\dagger_{b,a} (T_1(g))_{a,r} = \int dg \, \delta_{b,r} = \delta_{b,r}. \quad (C.57)$$

Putting everything together, we obtain the orthonormality relations:

$$\int dg \, T^{(j)}(g)_{q,r} T^{(j')}(g)^*_{a,b} = 1/\mathrm{Dim}(j)\delta_{j,j'}\delta_{q,a}\delta_{r,b}. \qquad (C.58)$$

With the parameterisation given earlier, the invariant measure of $SU(2)$ is written [7]:

$$dg = (constant)\sin(\frac{\alpha}{2})d\alpha \, d\cos\theta \, d\phi. \qquad (C.59)$$

$(0 \le \alpha \le 2\pi; 0 \le \theta \le \pi; 0 \le \phi \le 2\pi)$. We leave it to the reader to verify the orthonormality relations on the matrix elements $\mathcal{D}^{(1/2)}$ given in (C.28).

C.11   THE WIGNER–ECKART THEOREM

A very important case is that of a set of operators in the Hilbert space of a quantum system which transform according to an irreducible representation of the group:

$$U(g)\Phi^{(j)}_m U(g)^\dagger = T^{(j)}_{m,m'} \Phi^{(j)}_{m'}, \qquad (C.60)$$

where $U(g)$ are the unitary operators which represent the action of the transformations on the states of the system.

As before, we assume that $U(g)$ are completely reducible into finite-dimensional blocks, equation (C.32). We consider the matrix element of Φ between the states that belong to the representation k and l in the reduction of $U(g)$ (these representations are fixed and therefore it is not necessary to indicate the values of the other quantum numbers, α, β, which identify them within the decomposition of $U(g)$). We continue by inserting in the matrix element of the operator $1 = U(g)^\dagger U(g)$ (we always sum over repeated indices):

$$< l, s|\Phi_m^{(j)}|k, n > = < l, s|U(g)^\dagger U(g)\Phi_m^{(j)}U(g)^\dagger U(g)|k, n > =$$

$$= (T^{(l)})^*_{s,s'} < l, s'|\Phi_{m'}^{(j)}|k, n' > (T^{(k)})_{n,n'}(T^{(j)})_{m,m'}. \quad (C.61)$$

We now use the expression which gives the product of two matrices in terms of the irreducible components, via the Clebsch–Gordan coefficients, equation (C.37). We find

$$< l, s|\Phi_m^{(j)}|k, n > = \sum_{s',m',n'} T^{(l)}(g)^*_{s,s'} < l, s'|\Phi_{m'}^{(j)}|k, n' > \cdot$$

$$\cdot \sum_{J,\alpha,M} T^{(J)}(g)_{M,M'} C(k, n; j, m \,|J, M, \alpha) C^*(k, n'; j, m' \,|J, M', \alpha). \quad (C.62)$$

We can now integrate both sides on the group and use the orthonormality relations found in the previous section. This selects the representation with $J = l$ and fixes $M = s$, $M' = s'$ in the Clebsch–Gordan series. We find:

$$< l, s|\Phi_m^{(j)}|k, n > =$$

$$= \sum_{m',n',\alpha,s'} < l, s'|\Phi_{m'}^{(j)}|k, n' > C(k, n; j, m|l, s, \alpha)C^*(k, n'; j, m'|l, s', \alpha) =$$

$$= \sum_\alpha C(k, n; j, m|l, s, \alpha) \sum_{m',n',s'} C^*(k, n'; j, m'|l, s', \alpha) < l, s'|\Phi_{m'}^{(j)}|k, n' > =$$

$$= \sum_\alpha C(k, n; j, m|l, s, \alpha) < l||\Phi^{(j)}||k >_\alpha; \quad (C.63)$$

$$< l||\Phi^{(j)}||k >_\alpha = \sum_{n',m',s'} C^*(k, n'; j, m'|l, s', \alpha) < l, s'|\Phi_{m'}^{(j)}|k, n' > . \quad (C.64)$$

The final line of (C.63) represents the final result: the so-called *Wigner–Eckart theorem*. The matrix element can be non-zero only if the representation $T^{(l)}$ is present in the tensor product $T^{(j)} \otimes T^{(k)}$. The dependence of the matrix element on the magnetic quantum numbers is provided by the relevant Clebsch–Gordan coefficients multiplied by two numbers, the *reduced matrix elements*, (C.64), which depend exclusively on the representations j, k and l. The reduced matrix elements are equal in number to the number of times that the representation l appears in the reduction of the product $T^{(j)} \otimes T^{(k)}$.

Bibliography

[1] L. Maiani, and O. Benhar, *Relativistic Quantum Mechanics: an Introduction to Quantum Fields.* Taylor and Francis, to appear.

[2] F. Mandl, and G. Shaw, *Quantum Field Theory.* Wiley, 1984.

[3] N. Cabibbo, L. Maiani, and O. Benhar, *Introduction to Gauge Theories.* Taylor and Francis, to appear.

[4] L.D. Landau, and E.M. Lifshitz, *Quantum Mechanics: Non-Relativistic Theory*, 3rd ed., Vol. 3. Butterworth-Heinemann, 1977.

[5] C. Amsler *et al.* [Particle Data Group Collaboration], Phys. Lett. B **667** (2008): 1.

[6] L. O'Raifeartaigh, *The Dawning of Gauge Theory.* Princeton University Press, 1997.

[7] H. Weyl, *The Theory of Groups and Quantum Mechanics.* Dover Books on Mathematics, 1950.

[8] C. N. Yang, and R. Mills, Phys. **96** 191(1954).

[9] For the case of the double quantum well, see R. P. Feynman, R. B. Leighton, and M. L. Sands, *The Feynman Lectures on Physics.* Addison Wesley Longman (1970).

[10] S. L. Glashow, Nucl. Phys. **22** (1961): 579.

[11] F. Englert, and R. Brout, Phys. Rev. Lett. **13** (1964): 321.

[12] P. W. Higgs, Phys. Lett. **12** (1964): 132; Phys. Rev. Lett. **13** (1964): 508.

[13] S. Weinberg, *Phys. Rev. Lett.* **19**, 1264 (1967); A. Salam, in N. Svartholm: Elementary Particle Theory, *Proceedings of the Nobel Symposium 1968.* Lerum, Sweden (1968): 367-377.

[14] K. Winter, ed., *Neutrino Physics*, Cambridge Monographs on Particle Physics, Cambridge University Press, 1991.

[15] S. Schael *et al.* [ALEPH and DELPHI and L3 and OPAL and SLD and LEP Electroweak Working Group and SLD Electroweak Group and SLD Heavy Flavour Group Collaborations], Phys. Rept. **427** (2006): 257.

[16] See for example M. Gell–Mann, and Y. Ne'eman, *The Eightfold Way*. W. A. Benjamin, New York, Amsterdam, 1964.

[17] M. Gell–Mann, Phys. Lett. **8** (1964): 214-215.

[18] G. Zweig, *An SU(3) Model For Strong Interaction Symmetry And Its Breaking. 2*, CERN-TH-412.

[19] M. Y. Han, and Y. Nambu, Phys. Rev. **139** 4B, B1006 (1965).

[20] C. Bouchiat, J. Iliopoulos, and P. Meyer, Phys. Lett. **B38** (1972): 519.

[21] D. J. Gross and F. Wilczek, Phys. Rev. D **8** (1973): 3633; Phys. Rev. D **9** (1974): 980.

[22] H. D. Politzer, Phys. Rev. Lett. **30** (1973): 1346.

[23] Especially by G. Altarelli, and G. Parisi, Nucl.Phys. **B126** (1977): 298; Y.L. Dokshitzer, Sov. Phys. J.E.T.P. **46** (1977): 691.

[24] G. Altarelli, *The Development of Perturbative QCD*. Singapore, World Scientific (1994).

[25] H. Fritzsch and M. Gell–Mann, Proceedings of the XVI International Conference on High Chicago 1972 p.135 (J. D. Jackson, A. Roberts, eds.), eConf C **720906V2** (1972) 135 [hep-ph/0208010].

[26] S. Weinberg, Phys. Rev. Lett. **31** (1973): 494.

[27] See for example J. Erler, P. Langacker in J. Beringer et al. (Particle Data Group), Phys. Rev. **D**86, 010001 (2012).

[28] N. Cabibbo, Phys. Rev. Lett., **10**, 531 (1963).

[29] S. L. Glashow, J. Iliopoulos, and L. Maiani, Phys.Rev., **D2**, 1285 (1970).

[30] B. L. Ioffe and E. P. Shabalin, Yadern. Fiz. **6**, 828 [Soviet J. Nucl. Phys. **6**, 603 (1968)].

[31] M. Kobayashi, and T. Maskawa, Progr. Theor. Phys., **49**, 652 (1973).

[32] L. Wolfenstein, Phys. Rev. Lett. **51** (1983): 1945.

[33] A. Ceccucci, Z. Ligeti, and Y. Sakai, *The CKM quark-mixing matrix*, in [5].

[34] A. J. Buras, M. E. Lautenbacher, and G. Ostermaier, Phys. Rev. D **50** (1994): 3433. See also [35].

[35] D. P. Kirkby, and Y. Nir, *CP-violation in meson decays*, in [5].

[36] Y. Nambu, and G. Jona-Lasinio, Phys. Rev. **122**, 345 (1961).

[37] M. Gell–Mann, R. Oakes, and J. Renner, Phys. Rev. **175**, 2195 (1968).

[38] S. L. Glashow and S. Weinberg, Phys. Rev. Lett. **20**, 224 (1968).

[39] N. Cabibbo, and L. Maiani, *Weak interactions and the breaking of hadron symmetries*, Evolution of particle physics. Academic Press (1970): 50.

[40] L. Maiani, *CP and CPT violation in neutral kaon decays*, The Second DAPHNE Physics Handbook, Vol. 1, L. Maiani, G. Pancheri, and N. Paver (eds.), Frascati INFN (1996).

[41] M. Antonelli, and G. D'Ambrosio, *CPT Invariance Tests in Neutral Kaon Decay*, in [5].

[42] G. Gabrielse, A. Khabbaz, D. S. Hall, C. Heimann, H. Kalinowsky, and W. Jhe, Phys. Rev. Lett. **82** (1999): 3198.

[43] M. Hori *et al.* [ASACUSA Collaboration], Nature **475** (2011): 484.

[44] L. Wolfenstein, Phys. Rev. Lett. **13** (1964): 562.

[45] G. Isidori, Y. Nir, and G. Perez, Ann. Rev. Nucl. Part. Sci. **60** (2010): 355.

[46] T. Inami, and C. S. Lim, Prog. Theor. Phys. **65** (1981): 297 [Erratum-ibid. **65** (1981): 1772].

[47] D. J. Antonio *et al.* [RBC and UKQCD Collaborations], Phys. Rev. Lett. **100** (2008): 032001; C. Aubin, J. Laiho and R. S. Van de Water, Phys. Rev. D **81** (2010): 014507.

[48] J. L. Rosner and S. Stone, *Decay Constants of Charged Pseudoscalar Mesons*, in [5].

[49] M. Okamoto, PoS LAT **2005** (2006) 013 [hep-lat/0510113].

[50] O. Schneider, B_0–$\bar{B}^0$ *Mixing*, in [5].

[51] M. Bona *et al.* [UTfit Collaboration], JHEP **0803** (2008): 049.

[52] S. Herrlich, and U. Nierste, Nucl. Phys. B **476** (1996): 27.

[53] A. F. Falk, Y. Grossman, Z. Ligeti, and A. A. Petrov, Phys. Rev. D **65** (2002): 054034; A. F. Falk, Y. Grossman, Z. Ligeti, Y. Nir, and A. A. Petrov, Phys. Rev. D **69** (2004): 114021.

[54] A. Djouadi, J. Kalinowski, and M. Spira, Comput. Phys. Commun. **108**, 56 (1998) [hep-ph/9704448].

[55] J. F. Gunion, H. E. Haber, G. L. Kane, and S. Dawson, Front. Phys. **80** (2000): 1.

[56] A. Djouadi, Phys. Rep. **457** (2008): 1.

[57] J. R. Ellis, M. K. Gaillard, D. V. Nanopoulos, Nucl. Phys. **B106**, 292 (1976).

[58] M. A. Shifman, A. I. Vainshtein, M. B. Voloshin, and V. I. Zakharov, Sov. J. Nucl. Phys. **30**, 711 (1979).

[59] W. J. Marciano, C. Zhang, and S. Willenbrock, Phys. Rev. D **85** (2012): 013002.

[60] M. Carena, I. Low, and C. E. M. Wagner, JHEP **1208** (2012): 060.

[61] M. M. Kado, and C. G. Tully, Ann. Rev. Nucl. Part. Sci. **52** (2002): 65.

[62] L. Maiani, A. D. Polosa, and V. Riquer, New J. Phys. **14** (2012): 073029.

[63] LHC Higgs Cross Sections Working Group, https://twiki.cern.ch/twiki/bin/view/LHCPhysics/CrossSections.

[64] G. Aad *et al.* [ATLAS Collaboration], Phys. Lett. B **716**, (2012): 1.

[65] S. Chatrchyan *et al.* [CMS Collaboration], Phys. Lett. B **716**, (2012): 30.

[66] Tevatron New Physics Higgs Working Group and CDF and D0 Collaborations, arXiv:1207.0449 [hep-ex].

[67] N. Cabibbo, and R. Gatto, Phys. Rev. **116**, (1959): 1134; N. Cabibbo, R. Gatto, and C. Zemach, Il Nuovo Cimento **16** (1960): 168; G. Feinberg, P. Kabir, and S. Weinberg, Phys. Rev. Letters **3** (1959): 527.

[68] F. Peter, and H. Weyl, Math. Ann. **97** 737 (1927).

[69] G. De Franceschi, and L. Maiani, Fortschritte der Physik **18** (1965): 279.

[70] S. Coleman, *Aspects of Symmetry: Selected Erice Lectures.* Cambridge University Press, 1985.

[71] J. J. de Swart, Rev. Mod. Phys. **35** (1963): 916 [Erratum-ibid. **37** (1965): 326].

Index

Printed in the United States
by Baker & Taylor Publisher Services